Drought Management and Planning for Water Resources

T0174251

Drought Management and Planning for Water Resources

Edited by
Joaquín Andreu
Giuseppe Rossi
Federico Vagliasindi
Alicia Vela

CRC Press
Taylor & Francis Group
Boca Raton London New York

CRC Press is an imprint of the
Taylor & Francis Group, an **informa** business

CRC Press
Taylor & Francis Group
6000 Broken Sound Parkway NW, Suite 300
Boca Raton, FL 33487-2742

First issued in paperback 2019

© 2006 by Taylor & Francis Group, LLC
CRC Press is an imprint of Taylor & Francis Group, an Informa business

No claim to original U.S. Government works

ISBN-13: 978-1-56670-672-8 (hbk)
ISBN-13: 978-0-367-39190-4 (pbk)

Library of Congress Card Number 2005050550

Library of Congress Cataloging-in-Publication Data

Drought management and planning for water resources /edited by Joaquin Andreu ... [et al.].
    p. cm.
  Includes bibliographical references and index.
  ISBN 1-56670-672-6
  1. Water-supply--Management. 2. Droughts--Management. I. Andreu, J. (Joaquin)

TD345.D76 2005
363.6¢1--dc22
                                                 2005050550

Visit the Taylor & Francis Web site at
http://www.taylorandfrancis.com

and the CRC Press Web site at
http://www.crcpress.com

Preface

Water resources management in the arid and semiarid areas is a complex task, involving a large number of hydrologic, environmental, and management factors that have to be considered in order to supply sufficient water and to ensure the minimum levels of environmental protection and quality of life. Droughts, so frequent in the semiarid areas, intensify these problems even more. Since they are unpredictable phenomena (both in their occurrence and duration), prevision and preparation against droughts are key elements for minimizing their impact.

These circumstances have driven researchers to invest an important effort in the study of alternative, nonconventional means for obtaining water in prevision of drought periods, such as wastewater treatments, desalinization, or exploitation of deep groundwater, as well as the development of tools and strategies for conjunctive management and water saving that allow for optimizing the water resources management and preventing the scarcity periods.

The WAMME project (Water Resources Management Under Drought Conditions: Criteria and Tools for Conjunctive Use of Conventional and Marginal Waters in Mediterranean Regions) has investigated these subjects and applied the obtained methodologies and results to a series of study cases located in representative basins of the Mediterranean area. The objective of this book is to present these results to the potential users and the members of the international scientific community.

About the Editors

Joaquín Andreu, Ph.D., is a professor of hydraulic engineering at the University Politécnica de Valencia and presently technical director at the Júcar River Basin Agency (Confederación Hidrográfica del Júcar). He graduated with a degree in civil engineering (1977), obtained his MSc in civil engineering at the Colorado State University (1982), and obtained a Ph.D. in civil engineering (1984) at the University Politécnica de Valencia.

Since 1979 he has been developing a researching and teaching career at the University Politécnica de Valencia. He became a professor of hydraulic engineering in 1993, and from 2001 to 2004 he was the director of the Institute of Water Engineering and Environment at the University Politécnica de Valencia.

His main research interests are decision support systems for water management, and he has conducted several research projects (funded by the European Union, the Spanish Ministries of Science and Development and Foreign Affairs, and other European and international institutions) and directed 11 doctoral theses. He is also author or coauthor of 11 books and has published more than 40 papers and contributions to several research publications.

Giuseppe Rossi, Ph.D., is a professor of hydrology and water resources at the University of Catania, College of Engineering, Italy. He graduated with a degree in civil engineering from the University of Palermo. He was a research associate at the Institute of Hydraulics, Hydrology and Water Management, University of Catania (1971–1979). He was a visiting scientist at Colorado State University, Fort Collins (1977–1978) and at Universidad Politécnica de Valencia, Spain (2003–2004).

His main research interests are stochastic hydrology, analysis of hydrologic extremes (floods and droughts), and models for water resources management.

He was a partner or coordinator of several research projects on drought analysis and mitigation funded by the European Union (CEE-EPOCH, INCO-DC, INCO-MED, and MEDA programs) and coordinator on behalf of the University of Catania for several projects on water resources systems planning and operation funded by the Italian Ministry for University and Research and National Research Council.

He is the author of over 130 papers and editor of seven books in the fields of floods, droughts, and water resources. He is a member of the editorial boards of *Water Resources Management, International Journal of Water,* and *L'Acqua, Journal of the Associazione Idrotecnica Italiana.*

Professor Rossi is a member of several international associations (Hydraulics Research IAHR, Water Resources IWRA, EWRA, Water History IWHA, Irrigation and Drainage ICID), of Associazione Idrotecnica Italiana, Indian Association of Hydrologists, and Zelanti and Gioenia Academies).

Federico G. A. Vagliasindi, Ph.D., is a professor of environmental and sanitary engineering at the Department of Civil and Environmental Engineering, University of Catania (Italy). He received his MSc from Colorado State University, Fort Collins, and his Ph.D. from University of Washington, Seattle. He was a research associate at University of Salerno, Italy (1992), a research assistant at University of Washington, Seattle (1994–1997), and an associate professor at University of Catania, Italy (1998–2002).

Professor Vagliasindi's main research interests are in the fields of water and wastewater treatment and reuse, integrated solid waste management, and contaminated site characterization and remediation. He has participated in research projects funded by the European Commission, AWWARF (American Water Works Association Research Foundation, USA), EPRI (Energy and Power Research Institute, USA), Italian Ministry for University and Research, and Italian National Research Council. He has authored over 50 papers for journals or proceedings. He is coauthor of an AWWA research foundation manual on arsenic removal, and coauthor of the wastewater treatment chapter of the Italian *Handbook of Civil Engineering.*

Professor Vagliasindi is a member of the American Association of Environmental Engineering Professors and the International Water Association. He is a member of the Environmental and Sanitary Section of the Italian Great Risks Commission.

Alicia Vela, Ph.D., geologist, is currently a private consultant on hydrology, hydrogeology, and GIS applications to hydrology. She received her Ph.D. from the University Complutense de Madrid (1999), and worked as a researcher at the Remote Sensing and GIS Section of the University of Castilla–La Mancha (1999–2002), and at the Institute of Water and Environmental Engineering of the Universidad Politécnica de Valencia (2002–2003).

Dr. Vela has focused her research on remote sensing and GIS applications to hydrology, and mainly toward the development of GIS applications for modeling soil-water, soil-atmosphere, and soil-subsoil processes. She has participated in several research projects funded by the European Commission, the European Space Agency, and the Spanish Ministry for Science and Technology. She has authored or coauthored 20 papers for journals, proceedings, or monographies.

Dr. Vela is a member of the International Association of Hydrogeologists.

Contributors

Joaquín Andreu
Universidad Politécnica de Valencia
Instituto de Ingeniería del
 Agua y Medio Ambiente
Valencia, Spain

A. Cancelliere
University of Catania
Department of Civil and
 Environmental Engineering
Catania, Italy

Teodoro Estrela
Confederación Hidrográfica
 del Júcar
Valencia, Spain

Aránzazu Fidalgo
Confederación Hidrográfica
 del Júcar
Valencia, Spain

G. Giuliano
University of Catania
Department of Civil and
 Environmental Engineering
Catania, Italy

Daniel P. Loucks
Cornell University
Civil and Environmental
 Engineering
Ithaca, New York

Giuseppe Mancini
University of Catania
Department of Civil and
 Environmental Engineering
Catania, Italy

Javier Paredes
Universidad Politécnica de Valencia
Dpto. de Ingeniería Hidráulica y
 Medio Ambiente
Valencia, Spain

Miguel Angel Pérez
Universidad Politécnica de Valencia
Instituto de Ingeniería del
 Agua y Medio Ambiente
Valencia, Spain

Javier Ferrer Polo
Confederación Hidrográfica
 del Júcar
Valencia, Spain

Paolo Roccaro
University of Catania
Department of Civil and
 Environmental Engineering
Catania, Italy

Giuseppe Rossi
University of Catania
Department of Civil and
 Environmental Engineering
Catania, Italy

Andres Sahuquillo
Universidad Politécnica de Valencia
Dpto. de Ingeniería Hidráulica y
  Medio Ambiente
Valencia, Spain

Francesca Salis
Dip. Ingegneria del Territorio
  Sezione Hidráulica
Piazza d'Armi
Cagliari, Italy

Giovanni M. Sechi
Dip. Ingegneria del Territorio
  Sezione Hidráulica
Piazza d'Armi
Cagliari, Italy

Vicente Serrano-Orts
COPUT — Servicio de
  Planificación
Valencia, Spain

Salvatore Sipala
University of Catania
Department of Civil and
  Environmental Engineering
Catania, Italy

A. Solera
Universidad Politécnica de Valencia
Dpto. de Ingeniería Hidráulica y
  Medio Ambiente
Valencia, Spain

Federico G. A. Vagliasindi
University of Catania
Department of Civil and
  Environmental Engineering
Catania, Italy

Paola Zuddas
Dip. Ingegneria del Territorio
  Sezione Hidráulica
Piazza d'Armi
Cagliari, Italy

Contents

chapter one

Water management in Mediterranean regions prone to drought: The Júcar Basin experience

Javier Ferrer Polo
Confederación Hidrográfica del Júcar, Valencia, Spain

Javier Paredes
Universidad Politécnica de Valencia, Spain

Contents

1.1   Introduction

It is in the Mediterranean basins where the scarcity of water, irregular hydrology, and great water demands cause droughts to have important economic, social, and environmental consequences. Drought concept is complex due to the subjectivity function of the field of study and the development of the system. Many authors have defined the drought concept (Dracup et al., 1980; Yevjevich et al., 1983; Easterling, 1988; Rossi et al., 1992; Wilhite, 2000). It could be defined as a significant circumstantial decrease of the hydrologic resources, during a timeframe sufficiently prolonged, that affects an extensive area and that has adverse socioeconomic consequences.

Drought's importance lies in its slow and progressive nature, which makes Basin managers deny the event until they are completely inside it. An added difficulty is the impossibility of identifying cycles or periodical events. For these reasons mitigation actions are not implemented until the situation is critical, which means that emergency actions are not always efficient.

The subjectivity of the concept bases in the necessity of establishing different factors to characterize the drought, such as duration, threshold of definition, type of effect considered, or the degree of the consequence is considered. This subjectivity carries into the consideration of different concepts of droughts:

- Meteorological drought: defined as the precipitation decrease, with respect to the regular regional value, during a specific timeframe.
- Agricultural drought or soil humidity shortage, which does not satisfy specific crop growing needs during a specific timeframe.
- Hydrologic drought: decrease in surface and groundwater availability, with respect to regular values, within a management system during a temporary timeframe.
- Socioeconomic drought: defined as the effects of water scarcity on people and on economic activity due to drought. Avoiding these effects or minimizing them is part of management success.

The encircling character of the drought phenomenon has consequently been reduced to a traditional point of view with emergency actions and extraordinary resources used only when facing a critical situation. This point of view has been followed in Spain during the most recent droughts. There is another option where drought management is included inside the planning process with the analysis of the risk and planning for drought events. Several authors have examined this approach (Wilhite and Wood, 1985; Dziegielewski, 1986, 2003; Easterling and Riebsame, 1987; Riebsame et al., 1990; Grig and Vlachos, 1989; Wilhite, 2000). To obtain this objective a watch alert system has to be active in the region with objective indicators and drought scenarios examined in the planning process. Moreover, there must be an appropriate legal and administration framework.

1.2 Current Spanish legal framework

Article 58 of the Refunded Text of the Spanish Water Law provides government with guidelines to measure the hydraulic public domain in order to fight exceptional situations of droughts. These measures carry the declaration of public interest of the constructions and the proceedings with the objective of surpassing water scarcity situations. This point of view is the traditional focus that considers droughts as emergency situations, rejecting the planning approach and its advantages. The *White Water Spanish Book* (MIMAM, 2000) concludes that the most efficient solution is not to expect the emergency situation for using groundwater, but planning and managing water resources systems in an optimum way, with special attention to drought times. It was not until 2001, in the National Hydrologic Plan (Law 10/ 2001, July 5), when the basis of drought planned management was established in the legal framework. Article 27 of the cited law provides three ways of acting against drought:

- The Environmental Ministry will establish a global system of hydrological indicators that are alert to these kinds of situations and works as a general reference to the basin agencies for states emergency situations.
- The statement of these situations will be accompanied with the activations of the Special Plan developed by the agency basin and its alert measurements, containing exploitation rules and measurements related to the use of the hydraulic public domain. By law it is mandatory to develop these plans within two years of the law's promulgation.
- Public administration is responsible for the supply of urban systems with more than 20,000 people will develop a Drought Emergency Plan. These plans have to be working within four years of the law's promulgation.

With this new legal framework four tools have been created in order to plan and manage droughts:

- The Basin Drought Special Plan must contain operation rules of the systems in scarcity situations, structural actions, and rules of use of hydraulic public domain.
- Drought indicators established by the Environmental Ministry allow the control, identification, and warning of droughts in basins.
- River Basin District Drought Indicators allow the same functions as the global ones.
- Finally, emergency plans of water supply systems for populations over 20,000 materialize the actions, provide for the droughts plans, and allow administrative cooperation.

1.3 Droughts in the Júcar Basin Agency

According to the Spanish Constitution, water competence relapses over the central government if the basin spreads into different autonomic communities. Júcar Basin Agency (JBA) manages the water resources and the hydraulic public domain of the territory defined in Article 17 of Royal Decree 650/1987, of May 8, where territories of the basin agencies and the hydrological plans are defined. The territory of the JBA has a surface of 42,988.6 km², and it spreads over four autonomic communities. Figure 1.1 is a map of the territory where several basins are grouped in nine exploitation systems. The total population is 4,420,878 with a high seasonal growth of more than 4.7 million due to tourism.

The major water demand is from the agricultural sector, with an irrigated surface over 400,000 ha, which is 80% of the total demand. Precipitation in JBA is characterized by a remarkable spatial and temporal variability, with an annual mean value of 500 mm/year. There is spatial variability in some areas such as near the Júcar and Cabriel rivers and within the Marina Alta system, where the mean precipitation is over 800 mm/year, and others such as the Vinalopó basin, where the annual mean precipitation is under 250 mm/year. This situation of spatial irregularity among different system

Figure 1.1 Location map of the territory of the Júcar Basin Agency.

conditions determines the grade of regional drought vulnerability. Otherwise, most precipitation series within the JBA show a high temporal variability as is the case near the head of the Júcar and Turia rivers, where since 1940 one wet period has occurred (1958–1977) and three dry ones (1978–1985, 1991–1995, and 1998–2002).

1.4   Indicators and watch alert systems in the JBA

Although hydrological drought indices are most commonly used, indicators within a basin or territory must also be monitored to evaluate all the different types of droughts. Rossi et al. (2003) present a detailed study on requests and conditions of the watch alert systems and compile different systems developed throughout the world. JBA uses pluviometric series, aquifer piezometric levels, impaired inflows, and storage volumes in aquifers and reservoirs. Installation of this indicator system has different stages:

- Establishing indicators by units of exploitation. The JBA's territory includes several exploitation systems with different behaviors for droughts. In the definition of the indicators the particularities of each area have to be taken into account.
- For each system a weighted indicator system is established in order to obtain representative numerical results of the drought situation. At the same time a task of validation of the indicators has been done.
- Another phase is the continuous follow-up of the indicators of each system and the periodical reports. Moreover, thresholds have to be defined to declare alert and drought situations. To define these thresholds, it is important to review the empirical knowledge and to optimize the exploitation systems in drought situations. Finally, the use of these indicators allows the evaluation of the intensity and importance of the droughts.

Currently 34 indicators have been used in the JBA, located as shown in Figure 1.2. They are distributed as follows:

- Seven pluviometers, where the indicator is the number of millimeters of accumulated rain in the past 12 months.
- Nine piezometric levels, where the indicator is the piezometric level measured in meters over the sea level.
- Nine appraised stations, where the mean flow (hm^3/month) in the last quarter is the chosen variable.
- Nine reservoir storages, where the indicator is the storage value (hm^3).

Previous indicators are estimated in a nondimensional way, with values between 0 and 1, taking into account maximum and minimum historical data and considering seasonal variation. Four thresholds are defined over

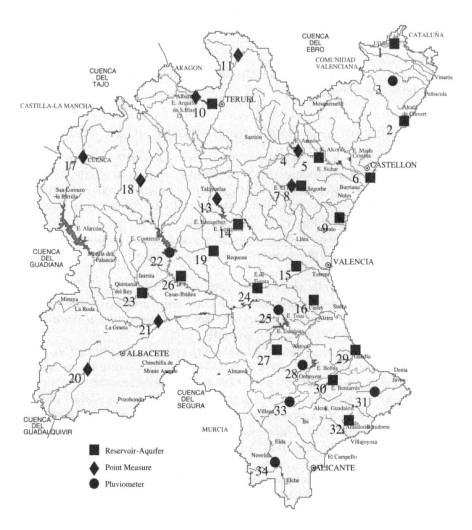

Figure 1.2 Drought indicators in the JBA.

the nondimensional indicators: stable situation (> 0.5), prealert (> 0.3), alert (> 0.15), and emergency (< 0.15), characterizing the state of each of the 34 indicators.

The use of weight factors, defined from the demand volume supply by the corresponding indicator, allows researchers to obtain mean values in the exploitation systems and to plot maps with the spatial distribution of the drought in the JBA, as shown in Figure 1.3.

Although the approach described here has a heuristic focus, it has been tested to be representative of the historic droughts and is used continuously in the management by the JBA. Consequently, the thresholds must be defined for each of these indicator levels.

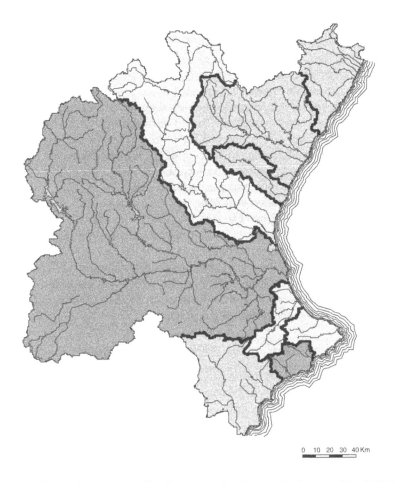

Figure 1.3 Example of spatial distribution of the drought indicators (April 2002).

1.5 Analysis of drought events in the JBA

In this section the two last and most important events of droughts in the JBA's territory are described.

1.5.1 Measures in the 1990–1995 drought

The drought that occurred in the first half of the 1990s was one of great impact of national proportion, although it had special influence in the south and east areas of the Iberian peninsula. There had been different national effects, where the most problematic was the cut in water supply during several hours each day for long periods in several important cities such as Granada, Jaen, Sevilla, Málaga, Toledo, Ciudad Real, and Puertollano. The crisis ended through different measures of infrastructures, such as transfers among basins and a wide use of groundwater and nonconventional resources using reclaimed wastewater.

Inside the JBA's territory the drought had economical, social, and environmental effects. The most serious effect was the problem of supply for Teruel city in the Turia system. Agricultural uses created problems in the traditional areas of the Júcar and Turia rivers, where the nonexistence of wells produced fragile superficial systems.

Several emergency measures, such as digging emergency wells, were taken by the Dirección General de Obras Hidráulicas (DGOH) and the JBA. Four wells were dug with measured flows of 280 l/s for supplying Teruel city. In the mentioned agricultural areas 447 l/s were measured in several drought wells. Figure 1.4 shows the location of the group of the wells dug by different administrations and particular users.

The government framework during that time consisted of complementary regulations. Particularly Royal Decrees 531/1992, from May 22, and 134/1994, from February 4, provide special administrative measures to manage the water resources according to Article 56 of the water law, which allowed the basin agencies, through their government meetings (Juntas de Gobierno), to constitute the permanent drought committee, which was instituted to reduce or eliminate demands and to force construction deposits, wells, or transport facilities when it is considered an emergency.

1.5.2   Measures taken in the 1998–2002 drought

The newest drought event in the JBA's territory occurred between 1998 and 2002, especially affecting Júcar, Marina Baja, and Vinalopó systems. Although it has been an extensive drought, anticipation and efficient management, with a reasonable use of all of the resources, have decreased the effects of this extreme situation.

Júcar system includes the basin of the river Júcar and its tributary Cabriel. It is the biggest basin of the JBA, at 2378 km^2 and with the highest water resources and demands. Because of the importance of the superficial supply and the capacity of the reservoirs, the main indicator that allows analyzing the evolution of the droughts in the Júcar system is the water storage at the three main reservoirs: Alarcón, Contreras, and Tous, the latter in operation since 1995. Figure 1.5 shows the annual and monthly evolution of this indicator in the past 20 years.

After the drought in the early 1990s the system greatly recovered its storage capacity in the year 1996 when the storage volume fluctuated between 700 and 1000 hm^3, which meant the greatest storage capacity since 1982. However, the dry hydrology in 1998 and 1999 reduced the storage volumes to less than 200 hm^3 in some months. Although in 2000 the system improved, due to an atypical hydrology in 2001, the storage volumes decreased less than 300 hm^3.

Marina Baja system is located in Alicante County and includes the basins of Algar and Amadorio rivers, with a total area of 583 km^2 and a Mediterranean semiarid climate. The main population is located near the coast, with considerably seasonal increases of population, more than 225% due to tourism,

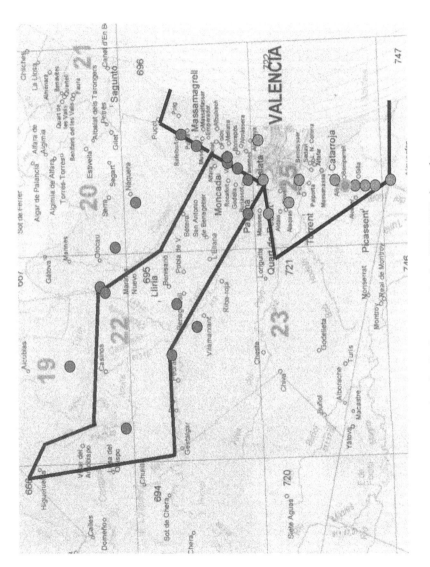

Figure 1.4 Wells constructed in the Turia system in the 1990–1995 drought.

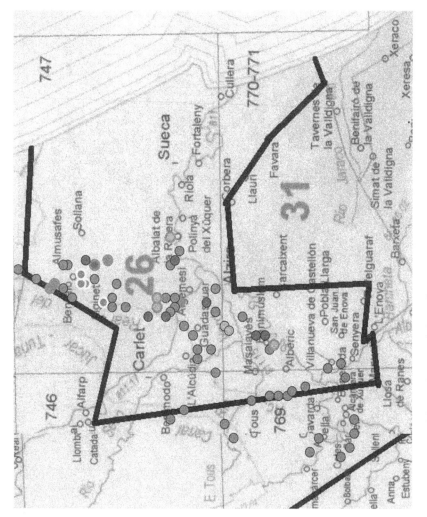

Figure 1.5 Monthly evolution of the volume storied in the Júcar system.

especially in Benidorm and Villajoyosa. Urban demand consumes almost all the water resources, which allows the reuse of the reclaimed wastewater for agricultural demands. Water supply is maintained by the Consorcio de Abastecimiento de la Marina Baja, which has a complex system of sub- and superficial water resources, including wells, regulation of the Algar spring pumping groundwater, and superficial storage in te Amadorio and Guadalest reservoirs. The storage volume of the two reservoirs is one of the most significant indicators in the drought analysis. Figure 1.6 shows monthly and annual evolution of this indicator. As can be determined in Figure 1.6, after the 1995 drought the system improved in 1998 with values over the averages but fell at the end of that year and maintained that situation of minimum values until October 2001. Water storage volume in that period was less than during the 1995 drought, although in the year 2003 it improved.

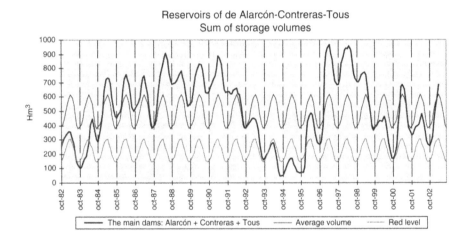

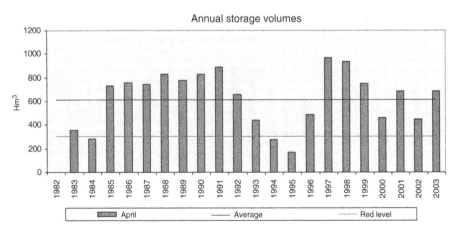

Figure 1.6 Monthly evolution of the volume of the Amadorio and Guadalest reservoirs.

Finally, the Vinalopó–Alacantí system is the southernmost one in all of the JBA; it includes the Monegre river, Rambuchar creek, and Vinalopó basin, with a total area of 2,786 km² and a mean annual precipitation of 320 mm/year, which characterizes the area as semiarid. Groundwater resources are overexploited, and the urban water is supplied to cities such as Alicante, Elche, San Vicente del Raspeig, Aspe, and Santa Pola by a consortium called Mancomunidad del Canal del Taibilla (MCT), which has external resources from the JBA. MCT has as its main objective supplying 76 urban demands located in three counties (Albacete, Alicante, and Murcia) of three different autonomous communities with a population of 1,800,000 people and a seasonal increase of 700,000 people. Three different origins supply the system: wells of the system, superficial resources from the Taibilla river, and resources from the Tajo–Segura Transfer (ATS), limited by law to 130 hm³/year. As can be see in Figure 1.7 the decrease of flows in the Taibilla river is very significant in the 1994–1996 period until October 1998, where there is a

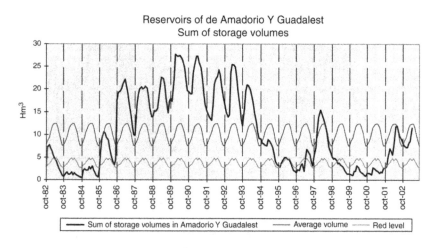

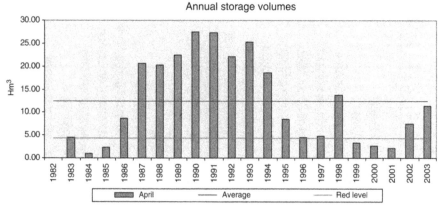

Figure 1.7 Monthly inflows of Taibilla river.

further decreased tendency finalizing in a historic minimum situation, 40–45 hm³, with this situation still remaining.

After characterizing the droughts, it is interesting to analyze the measures comparing the situation in 1995, where the measures consisted of incrementing groundwater resources use, and the last event, where the measures consisted of the conjunctive management of the system with both groundwater and surface resource uses taking advantage of the transfers among systems. The problem has been dealt with in a conjunctive way for the three systems due to the link among them. This connection is possible because of the use of three different facilities: ATS channel, MCT water supply system, and the Fenollar–Amadorio channel.

Moreover, the ATS channel is unique in that the Alarcón reservoir, in the Júcar system, is used in passing. This allows the use of water from the Júcar system in order to transfer it to the MCT system. On the other hand the Fenollar–Amadorio channel was developed as an emergency measure in the 1995 drought, and its purpose was to connect the MCT system with the Marina Baja water supply system. Figure 1.8 shows the connecting facilities among the three systems.

Two things were accomplished as a result of the 1998–2002 drought: the transfer of water from the Júcar system to the Marina Baja and Vinalopó–Alacantía systems and the creation of drought wells in the Júcar system. Actually it is an optimization of the water resources since the withdrawal of water from the Júcar system is substituted by an increment of the groundwater

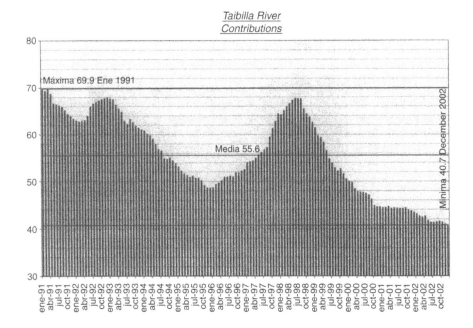

Figure 1.8 Transfer facilities among the Júcar, Marina Baja, and Vinalopó systems.

Table 1.1 Water Transfers from Júcar System in the 1998–2002 Period

Year	Receiver	Volume (hm³)		
		Origin	End	Effective
1999	Marina Baja	5.5 + 3.6	4.7 + 3.5	9.1
2000	Marina Baja	8.8	7.5	8.8
2001	Marina Baja	11.7	10.0	11.7
2002	Marina Baja	4.1	3.5	0.0
2000	M.C. Taibilla	2.0	1.8	2.0
2001	M.C. Taibilla	4.5	4.0	4.5
2002	M.C. Taibilla	10.9	10.0	10.9

resource. The transfer carried out in this period can be seen in Table 1.1, which compares the quantity of derived water and the final transfer of water due to losses along the way. The total water transferred in this period was 47 hm³.

These transfers have been complemented in 2002 by a group of 24 drought wells located in the low basin area of the Júcar river, creating an instantaneous total flow of 2.593 l/s. Many of these drought wells had been created during the 1990–1995 drought period.

There are some differences between the drought events of the mid to late 1990s that affect three aspects of drought management: facilities, legal framework, and administration measures.

1.5.2.1   Facilities
In the last event the Fenollar–Amadorio emergency channel, developed after the previous drought event, was available.

1.5.2.2   Legal framework
The legal framework has changed since 1995, with the development of legal tools that improve the management of the droughts in the JBA: Hydrologic Júcar Basin Plan (HJBP) and the change of the ATS management law.

First, the HJBP, published by Royal Decree 1664/1998 of July 20, has settled three points that have allowed the use of surplus water in the Júcar system with the objective of decreasing the water scarcity problems in other systems:

- A maximum of 80 hm³ is established to transfer water to Vinalopó and Marina Baja systems.
- Reserves set in the HJPB can be used to improve the environmental deficit or to decrease temporal problems in a water supply, while the concessions are not materialized.
- The use of the Alarcón reservoir in order to optimize the management of all the systems must be provided in a specific agreement between the Unión Sindical de Usuarios del Júcar (USUJ), users who promoted the construction of the reservoir, and the Environmental Ministry.

Table 1.2 Reserve Curve Established in the Alarcón Agreement

Month	Oct	Nov	Dec	Jan	Feb	Mar	Apr	May	Jun	Jul	Aug	Sep
Vhm³	278	287	287	326	334	326	311	278	263	263	263	263

This agreement to use the Alarcón reservoir in the unitary and optimized management of the Júcar system was published July 23, 2001, by USUJ and the Environmental Ministry, establishing a reserve curve, as shown in Table 1.2, constituted by storage volumes of the Alarcón reservoir with the objective of guaranteeing the USUJ rights. As the agreement says, "The indicated volumes from the regulation of the system will be reserved only to USUJ members, considering useful and available volumes of each reservoir of the system."

However, the agreement allows for the use of the resources of the Alarcón reservoir to other users if they finance the substitution of superficial use of groundwater resources to the USUJ users:

> If due to some circumstances, the Basin Agency, consulted the Withdrawal Committee, resolves the use of resources from the Alarcón Reservoir or other resources stored from USUJ users when the stored volume was not higher than the indicated in the previous table, the beneficiaries users without right to the mentioned water must pay to USUJ the total cost of the substitution of the volumes obtained from the USUJ area groundwater or from whichever source.

With this objective the agreement specifies a compensatory price $(1/m^3)$ by agreement or by arbitration of the Basin Agency: "In these cases, and previously to the execution of the measure, it will be settled the compensation, function of m^3, by users agreement. In case of disagreement the compensation will be settled by the Basin Agency, taking into account the parts, with a reasoned resolution."

Second, the use of the ATS facility, with a different objective from the original one, has been possibly due to Royal Decree Law 8/1999, May 7, which modifies Article 10 of the Law 52/1980, October 16, of the economic regime regulation of the management of the Tajo Segura channel. This modification allowed the use of the facility by different users than the original ones: "Independently of the previous articles, the uses with self resources of the Segura, Sur, and Júcar basins, foreseen in the Hydrologic Basin Plans, can use this facility to transfer and supply water among places inside the same hydrologic planning territory, paying a rate result of applying the approach established in the article 7."

1.5.2.3 *Complementary administrative measures*
To materialize these measures is an exhaustive and laborious administrative task, but quickly enough, after deliberations and agreements of the JBA Government Assembly, the resolution of the president authorized the transfer.

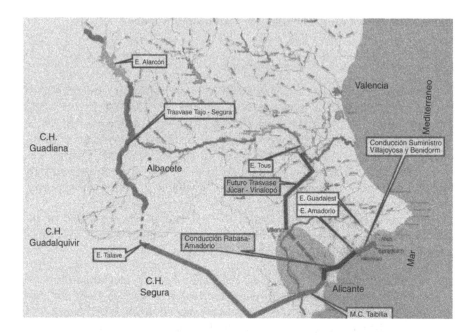

Figure 1.9 Facilities used in the last drought event.

Table 1.3 gives a detailed list of the administrative actions that made each transfer possible. In the same way, the transfer of 2002 and the digging of drought wells required additional authorization from the Basin Agency, taking into account the Withdrawal Committee.

Finally, an interesting issue is the economic quantification of the total cost of the extraordinary transfers. In the Marina Baja 2002 case it was 0.50426 €/m³, with the details shown in Table 1.4.

Table 1.3 Résume of the Administrative Actions

Year	Receiver	Initial volume (hm³)	Government assembly	Presidency resolution
1999	Marina Baja	5.5	12/5/1999	24/5/1999
2000	Marina Baja	8.8	4/5/2000	5/5/2000
2001	Marina Baja	11.7	21/2/2001	23/2/2001
2002	Marina Baja	4.1	24/4/2002	26/4/2002
2000	M.C.Taibilla	2.0	25/9/2000	26/9/2000
2001	M.C.Taibilla	4.5	26/7/2001	27/7/2001
2002	M.C.Taibilla	10.9*	17/6/2002	18/6/2002

*Drought wells. Withdrawal Committee 10/07/02.

Table 1.4 Example of the Economic Costs Associated with the 2002
Marina Baja Transfers

Concept	€/m³
Canon due to the use of Alarcón and Tous reservoirs in the Júcar Basin	0.020590
Canon due to the use of ATS facility	0.059882
Canon due to he use of MCT infrastructure	0.127797
Amortization of the Fenollar–Amadorio Channel	0.050399
Operation costs of the Fenollar–Amadorio Channel	0.203593
Cost due to the use of the drought wells	0.041999
Total	0.504260

1.6   Conclusion

The Mediterranean basins, with a spatial and temporal irregular hydrology
and a high use of water resources, are very vulnerable to drought events.
For this type of basin it is necessary to modify the traditional point of view
of emergency management and instead plan for a drought event. For this
reason it is important to develop a watch alert system using objective indicators of the drought events.

Two drought events have occurred in the JBA territory; one between 1991
and 1995, which was important because of the generalized consequences in the
entire territory, the second in 1998–2002, which influenced only three systems.
It is interesting to examine the conjunctive measures developed in the affected
exploitation systems. It has been possible, with the implementation of transfer
facilities, to increase the groundwater resources and global management of the
systems, under an adequate legal and economic framework, that takes into
account the rights over the resources and the infrastructure. Finally, the role of
the Basin Agency administration, with its quick actions according to the situation
and the type of the event, has been of great benefit to the region.

1.7   Acknowledgments

Thanks are due to Confederación Hidrográfica del Júcar for the information
supporting this chapter. The writing and editing have been funded by WAMME:
Water Resources Management Under Drought Conditions (within the European
INCO.MED Program), and SEDEMED: Secheresse and desertification dans le
bassin Mediterranee (within the European INTERREG program).

References

Dracup, J. A., Lee, K. S., and Paulson, E. G. (1980). On the definition of droughts.
  Water Resour. Res. 16(2), 297–302.
Dziegielewski, B. (1986). *Drought management options. Drought management and its
  impacts on public water systems.* Water Science and Technology Board, National
  Research Council. Washington, DC: National Academy Press.

Dziegielewski, B. (2003). Long-term and short-term measures for coping with drought. In G. Rossi, A. Cancelliere, L. S. Pereira, T. Oweis, M. Shatanawi, and A. Zairi (Eds.), *Tools for drought mitigation in Mediterranean regions*. Water Science and Technology Library.

Easterling, W. E., and Riebsame, W. E. (1987). Assessing adjustments in agriculture and water resources systems. In D. A. Wilhite, W. E. Easterling, and D. A. Wood (Eds.), *Planning for drought: Toward a reduction of social vulnerability*. Boulder, CO: Westview Press.

Easterling, W. E. (1988). Coping with drought hazard: Recent progress and research priorities. In F. Siccardi and R. L. Bras (Eds.), *Natural disasters in European Mediterranean countries* (pp. 231–270). Perugia, Italy: NSF and NRC.

Grig, N. S., and Vlachos, E. C. (1989). *Drought water management: Preparing and responding to drought*. Ft. Collins, CO: International School for Water Resources, Colorado State University.

MIMAM, 2000. *Libro Blanco del Agua en España. Secretaría de Estado de Aguas y Costas. Dirección General de Obras Hidráulicas y Calidad de las Aguas*. Ministerio de Medio Ambiente.

Riebsame, W., Changnon, S., and Karl, T. (1990). *Drought and natural resource management in the United States: Impacts and implications of the 1987–1990 drought*. Boulder, CO: Natural Hazards Research and Applications Information Center, University of Colorado.

Rossi, G., Cancelliere, A., Pereira, L. S., Oweis, T., Shatanawi, M., and Zairi, A. (Eds.). (2003). *Tools for drought mitigation in Mediterranean regions*. Water Science and Technology Library.

Rossi, G., Benedini, M., Tsakiris, G., and Giakoumakis, S. (1992). On regional drought estimation and analysis. *Water Resour. Manage.* 6, 249–227.

Yevjevich, V., Da Cunha, L., and Vlachos, E. (Eds.) (1983). *Coping with droughts*. Littleton, CO: Water Resources Publications.

Wilhite, D. A., and Wood, D. A. (1985). Planning for drought: The role of state government. *Water Resour. Bull.* 21(1), 31–38.

Wilhite, D. A. (2000). Preparing for a drought: A methodology. In D. A. Wilhite (Ed.), Drought. A global assessment Vol. II (pp. 89–104). London and New York: Routledge.

chapter two

Criteria for marginal water treatment and reuse under drought conditions

Giuseppe Mancini, Paolo Roccaro, Salvatore Sipala, and Federico G. A. Vagliasindi
University of Catania, Italy

Contents

2.1   Introduction

Scarcity of water in arid and semiarid regions causes development of appropriate plans, including both long- and short-term measures, to overcome the effects of drought events (Lazarova et al., 2001). Strategies to overcome the drought risk can be summarized in three main categories:

- Increase of the availability of resources, including non-conventional resources
- Education about water demands
- Minimization of drought impacts including appropriate operation rules of water supply systems

One of the most widely adopted measures, among the short-term ones, is the augmentation of the water supply by means of additional sources to increase robustness and resilience of the water system. These extra resources are often defined as unconventional or marginal waters, and can substitute intensively exploited conventional resources (e.g., fresh surface water and ground water) or can be used conjunctively to satisfy demand peaks or to cover water shortages during drought periods.

The term "marginal" is generally utilized to indicate water where the chemical, physical, and microbiological properties and its temporal and site availability are very specific, making its use unsafe, unreliable, and not productive unless it undergoes a special treatment (physical, chemical, or microbiological). Good quality water requiring high operational costs (deep ground water) can also be defined as marginal. Although there is no universal definition of marginal quality water, for all practical purposes it can be defined as water that possesses certain characteristics, which have the potential to cause problems when used for an intended purpose (FAO, 1992).

A not exhaustive list of the different categories of marginal water includes seawater and brackish water, domestic sewage water, irrigation drainage water, urban flood water, deep aquifer water, water found in remote areas whose exploitation requires high investment and high operational costs, and any other water that cannot be used directly in a safe beneficial manner.

An appropriate use of marginal waters requires a lot of cautions, either from an economic point of view but, above all, from the related environmental and sanitary implication (Anderson et al., 2001).

The specific objective of this work was to develop criteria for marginal water treatment and reuse under drought conditions, taking into account the minimum water quality requisites, the level of treatment and the related cost, and the hygienic constraint as a function of the final uses.

The main results obtained can be summarized as follows:

- A set of criteria and guidelines for marginal water quality and treatment as a function of its different uses
- A web-based information system (WBIS) to guide the screening and selection of the proper treatment for water reuse in each specific application

2.2   *Potential applications for marginal waters*

A partial remedy for water deficiencies occurring in arid and semiarid Mediterranean regions, especially when drought periods occur, is the recourse to marginal water resources, such as treated wastewater, saline or brackish waters, and deep ground waters. Several potential applications for these unconventional water resources are available, including:

- Agricultural irrigation (surface, sprinkler, and drip irrigation)
- Industrial applications (process water, cooling water, boiler-feed water)
- Urban dual distribution systems (one line for drinking water supply and the other for reclaimed wastewater) for subpotable uses (gardens irrigation, toilet flushing, etc.)
- Ground water recharge
- Wetland construction

Each application involves specific technical and hygienic issues.

2.2.1 Agricultural irrigation

Especially in arid and semiarid countries, where the lack of conventional water resources makes it difficult and expensive to ensure the total satisfaction of the water demands, it is necessary to take into serious consideration the possibility of using marginal water resources for irrigation. It is generally accepted that wastewater used in agriculture is justified from an agronomic and economic point of view, but care must be taken to minimize adverse health and environmental impacts. Particularly, in order to guarantee the public health safeguard and the environment protection, wastewaters reused for irrigation purposes need to reach different qualitative requisites depending on the specific applications and the select irrigation technique. The latter fall into three categories: surface, sprinkler, and drip irrigation.

Surface irrigation systems require less equipment than sprinkler systems and are not subject to spray drift problems. These irrigation systems are characterized by low capital costs but do not uniformly distribute the water on the soil layers. When surface irrigation is utilized, the farmers are in direct contact with the wastewater, causing notable risk for their health, especially if wastewater with inadequate quality is used.

The sprinkler irrigation can be implemented by several plant types and is suitable for all soil and crop typologies. This technique of irrigation, spreading the water on the land, determines a uniform distribution of water. With sprinkler irrigation, however, the contact between wastewater and irrigated crops is inevitable. One of the main health problems with this technique is the aerosols formation and the related risk for the workers and for people living close to the irrigation area. For this reason, reclaimed wastewater used in the spray irrigation must have good hygienic-sanitary characteristics, and an effective level of treatment has to be provided to reduce the risk of disease contraction. Barriers must be included in the field layout to minimize spray drift onto roads and dwellings.

Different studies have shown that the best irrigation technique for wastewaters reuse is the localized irrigation (drip irrigation, bubblers, microsprinklers, etc.), both subsurface and superficial. This specific technique, applying the water around each plant or group of plants and wetting the root zone only avoids the direct contact of wastewaters with the products and the agricultural operators. The irrigation of arboreal crops by localized irrigation would allow the use of partially treated wastewater, even with high bacterial content, therefore exploiting the high quantity of nutrients to increase soil fertility. However, localized irrigation causes significant technological problems due to the potential clogging of the microsprinklers, which can influence the functionality of the irrigation system.

Besides the irrigation technique, the required quality characteristics for the reclaimed wastewater depend on the type of irrigated crops. Specifically, three main types of cultivation, in order of health risk, can be considered: nonedible cultivation, edible cultivation after treatment, and directly edible cultivations. Obviously, the wastewater reused for the irrigation of direct edible cultivation could have optimal microbiological characteristics, in order to guarantee the protection of public health.

2.2.2 Ground water recharge

Ground water recharge with treated wastewater can be pursued in order to achieve the following:

- Contrast saltwater intrusion in coastal aquifers
- Provide further treatment for future reuse
- Augment potable or nonpotable aquifers
- Provide storage of reclaimed wastewater
- Control or prevent ground subsidence

Infiltration and percolation of reclaimed water take advantage of the subsoil's natural ability of biodegradation and filtration, thus providing additional *in situ* treatment of the wastewater and increasing the reliability of the overall wastewater management system. Depending on the method of recharge, hydrogeological conditions, and other factors, from the quality point of view, the treatment achieved in the subsurface layers may eliminate the need for expensive advanced wastewater treatments.

Ground water aquifers also constitute a natural reservoir, providing a free storage volume for the reclaimed wastewater. Irrigation demands are often seasonal, requiring large storage facilities and alternative means of disposal when reclaimed wastewaters are utilized but irrigation does not take place. Besides, suitable sites for surface storage facilities may not be available, economically feasible, or environmentally acceptable.

Although there are obvious advantages associated with ground water recharge, there are also possible disadvantages to consider:

- Extensive land areas may be needed for spreading basins
- Energy and injection wells for recharge may be prohibitively costly
- Recharge may increase the danger of aquifer contamination, and aquifer remediation is difficult, expensive, and may take years to be accomplished
- Not all added water may be recoverable
- The area required for operation and maintenance of a ground water supply system (including the ground water reservoir itself) is usually larger than that required for a surface water supply system
- Sudden increases in water supply demand may not be satisfied due to the slow movement of ground water

The quality of the water sources used for ground water recharge has a direct link with operational aspects of the recharge facilities and also with the allowed use of the recovered water. Generally, the main source water characteristics to be considered are suspended solids, dissolved gases, nutrients, biochemical oxygen demand, microorganisms, and the sodium adsorption ratio (which affects soil permeability). The constituents that have the greatest potential effects when potable reuse is expected include organic and inorganic toxicants, nitrogen compounds, and pathogens.

2.2.3   Industrial reuse

Many industries practice water recycling routinely, treating and using wastewater from one process in the same (recycle) or another process (reuse), one or more times. For example, many cooling towers, used in oil refineries and power generating plants relying on limited freshwater supplies, recycle water as many as eight times before discharging (blowing down) the concentrated brine to waste. Some industrial effluents are used for irrigation of landscaping or for process water at another industry. Industrial effluents can contain a large variety of pollutants such as heavy metals, toxic elements, and high content of organic matter. Where the cost of water is high enough, industries find it more economical to segregate the different wastewater streams and to treat and reuse water from different processes.

The industrial sector continuously requires large quantities of water. It is esteemed that around 25% of water demand in the world is correlated to industrial applications. In some heavily industrialized states in the U.S., industrial demand accounts for as much as 43% of the total.

In an industrial establishment water can be employed for different purposes, including: first matter, manufacture agent, energetic source to the liquid or vapor state, heat transfer, and other general uses (toilet flushing, irrigation, etc.).

Considering the large volume of water required in the industrial sector, the use of treated wastewater can be advantageous when the industries are located close to treatment plants serving strongly urbanized areas, in order to have a considerable treated flow. This managerial strategy could allow a notable savings of conventional water resources, which could be used for other applications.

As for economic convenience, it depends on many factors such as: the quality of available water, the additional treatments necessary for reaching the desired quality, and the distance from the point of use. Table 2.1 shows the industrial water reuse quality concerns and suitable treatment processes related to different contaminants.

2.2.4   Urban reuse

Marginal waters, and particularly treated wastewater, can be used in the urban areas for different nondrinkable purposes, such as:

- Irrigation of public parks and recreational centers, athletic fields, school yards and playing fields, highway medians and shoulders, and landscaped areas surrounding public building and facilities
- Irrigation of landscaped areas of single-family and multifamily residences and other maintenance activities

Table 2.1 Industrial Water Reuse Quality Concerns and Appropriate
Treatment Process

Parameter	Potential problem	Advanced treatment
Residual organics	Bacterial growth, slime/scale formation, foaming in boilers	Nitrification, carbon adsorption, ion exchange
Ammonia	Interferes with formation of free chlorine residual, causes stress corrosion in copper-based alloys, stimulates microbial growth	Nitrification, ion exchange, air stripping
Phosphorous	Scale formation, stimulates microbial growth	Chemical precipitation, ion exchange, biological phosphorous removal
Suspended solids	Deposition, "seed" for microbial growth	Filtration, microfiltration, ultrafiltration
Ca, Mg, Fe, and Si	Scale formation	Chemical softening, precipitation, ion exchange

Source: Adapted from U.S. EPA, *Guidelines for Water Reuse*, 1992.

- Irrigation of landscaped areas surrounding commercial, office, and industrial developments
- Irrigation of golf courses
- Commercial uses such as vehicle washing facilities, window washing, mixing water for pesticides, herbicides, and liquid fertilizers
- Ornamental landscape uses and decorative water features, such as fountains, reflecting pools, and waterfalls
- Dust control and concrete production on construction projects
- Fire protection
- Toilet flushing in commercial and industrial buildings

Urban reuse can include a vast range of possibilities, from the common residential uses to commercial and industrial. To reduce health hazards it is necessary to have dual distribution systems. In such distribution systems, reclaimed water is distributed to the various uses with a specific pipe network separated from the distribution network of drinking water. Some dual distribution systems have been operating since the 1970s in the U.S. Other urban reuse projects have been carried out in Japan and China. A pioneer project of urban wastewater reuse has been developed in the southern suburb of the city of Changzi, Shanxi Province of China. This project reused directly about 5000 m³/d of treated effluent (two-stage attached-ground biological treatment process, followed by sand filtration and disinfection) for washing, boiler supply, air pollution control, cooling, washroom flushing, and landscape irrigation.

2.2.5   Natural and manmade wetlands

Constructed wetlands (CW) are defined as "designed and man-made complex(es) of saturated substrates, emergent and submerged vegetation, animal life and water that simulates natural wetlands for human use and benefits" (Hammer, D.A. and Bastian, R.K., 1989). They have been used for wastewater treatment since the 1960s in Europe. Other names for constructed wetlands include rock reed filters, vegetated submerged beds, submerged bed flow systems, root zone systems, microbial rock filters, and hydrobotanical systems. CW are used for municipal wastewater treatment, acid mine drainage, industrial process water, agricultural point and nonpoint discharges, stormwater treatment or retention, and as a buffer zone to protect natural wetlands.

The advantages of constructed wetlands include inexpensive capital and maintenance costs, ease of maintenance, relative tolerance to changes in hydraulic and biological loads, and ecological benefits. Disadvantages include large land area requirements, lack of a consensus on design specifications, complex physical, biological, and chemical interactions providing treatment, pest problems, and topography and soil limitations.

Reclaimed wastewater can be used for creating wetlands in which flora and fauna can flourish, with particular reference to the creation or restoration of wet areas that constitute the natural habitat and the shelter for many animals and wild plants.

2.3   Issues in marginal waters utilization

The use of marginal water can cause several technical, economic, hygienic, and environmental problems, depending on the specific utilization (agricultural, industrial, urban, etc.) and the characteristic of available water (wastewater, brackish water, deep ground water, etc.). Table 2.2 shows a synthesis of the principal sanitary, technical, and hygienic problems that emerge from different specific applications of marginal water reuse.

2.3.1   Criteria for marginal waters utilization under drought conditions

2.3.1.1   Existing standards for water reuse in non-Mediterranean countries

Water reuse is well established in water-short regions of the U.S., Japan, and China, and it is receiving increased consideration in other parts of the world where traditional water supply sources are being stretched to their limits. Regulations and guidelines are being promulgated in many countries. The difference between regulations and guidelines is that regulations are enforceable by law, while guidelines are not legally enforceable, and compliance is voluntary. The water reclamation and reuse criteria in the U.S. are mainly based on health and environmental protection and principally regulate wastewater

Table 2.2 Technical and Hygienic-Sanitary Problems for Different Marginal Water Reuse Alternatives

Reuse alternative	Type of application	Problems
Agricultural	Superficial irrigation	Possible contact with cultivation; hygienic risks for the farmers; advanced treatments might reduce the concentration of nutrients; employment techniques less compatible with modern agricultural needs
	Sprinkler irrigation	Possible contact with cultivation; formation of aerosols; advanced treatments might reduce the concentration of the nutrients
	Drip irrigation	The use of only partially treated wastewater, with high nutrient content, increases the risk of soil porosity blockage
Industrial	Cooling water	Scaling or corrosion; biological growth caused by the presence of nutrients and organic material; obstruction due to deposits of particle material; production of aerosol and dangerous sprays for the workers
	Water for boilers	Scaling due to calcium and magnesium deposits; request for a high quality water
	Processing water	Function of the specific use (paper and cellulose, chemical and textile industry, etc.)
Urban	Toilet flushing, vehicle washing, fire protection system, etc.	Installation of a dual system for the distribution of treated wastewater, very expensive in the already developed urban areas; caution is required to prevent connection with the potable distribution net
Ground water recharge	Superficial spreading	Requirement of large infiltration basins; risk for ground water contamination; obstruction of the infiltration basins due to the formation of algae and particulate matter deposition; high operation and maintenance costs
	SAT (Soil Aquifer Treatment)	Necessity to use land which is hydrogeologically ideal for such practice
	Direct injection	Only feasible where ground water is shallow and well confined; obstruction can occur due to the accumulation of organic and inorganic solids; required characteristics for the reuse are similar to those for potable water
Environmental improvement	Constructed wetland	Risk of possible water contamination

treatment, reclaimed water quality, treatment reliability, distribution systems, and reuse area controls. California and Florida, which have several active reuse projects, have comprehensive regulations and prescribe restrictive requirements depending on the end use of the treated wastewater. The states that have not developed their criteria can make reference to published guidelines by the U.S. Environmental Protection Agency (EPA). This agency, in conjunction with the U.S. Agency for International Development, has published *Guidelines for Water Reuse* in 1992. The guidelines address all important aspects of water reuse including recommended treatment processes, reclaimed water quality limits, monitoring frequencies, setback distances, and other controls for various water reuse applications.

Guidelines for water reclamation and reuse are also provided by the World Health Organization (WHO). In 1985, a meeting of scientists and epidemiologists was held in Engelberg, Switzerland, to discuss the health risks associated with the use of wastewater for agriculture and aquaculture. The meeting results were confirmed by a WHO congress on Health Aspect of the Use of Treated Wastewater for Agriculture and Aquaculture held in Geneva in 1987. The final document was published by WHO as "Health Guidelines for the Use of Wastewater for Agriculture and Aquaculture." Table 2.3 shows a comparison of the microbiological quality guidelines and criteria for irrigation by WHO (1989), the U.S. EPA (1992), and the State of California (1978) (Asano and Levine, 1996).

2.3.1.2   Existing standards for water reuse
in Mediterranean countries

Many criteria and guidelines for the wastewater reclamation and reuse exist in the Mediterranean area countries. In Italy the general provisions on treated wastewater reuse were introduced by the Legislative Decree 152, May 11, 1999 (based on the EU directive 91/271), whereas specific regulations were promulgated with the Ministerial Decree 185, June 12, 2003. The new standards, not taking into account different agricultural reuse options and application techniques, are considered by operators and scientists as excessively restrictive. Furthermore, in order to cope with these standards, advanced treatments are required, which will result in high costs, often making the reuse of wastewater economically unfeasible.

In Spain the national water law (Ley de Aguas, 29/1985) introduced the basic conditions for the direct reuse of wastewaters according to the treatment processes, water quality, and accepted uses (there are no standards so far).

In Israel recent new criteria were adopted, based on a series of barriers that have to be met. The barriers are adjusted to the plants' characteristics, effluent quality, application method, harvesting practices, and timing of cultivation. These barriers are also adjusted to industrial utilizations and effluent disposal into public sites such as lakes, flowing streams and creeks, recreation reservoirs, and natural reserve sites. Effluent reuse in urban areas can be implemented for public garden irrigation, toilet flushing in public

Table 2.3 Comparison of the Microbiological Quality Guidelines and Criteria for Irrigation by the WHO (1989), the U.S. EPA (1992), and the State of California (1978)

Institution	Reuse conditions	Intestinal nematodes[a]	Fecal or total coliforms[b]	Wastewater treatment requirements
WHO	Irrigation of cereal crops, fodder crops, pasture, and trees	< 1/L	No standard recommended	Stabilization ponds with 8–10 day retention or equiv. removal
WHO	Irrigation of crops likely to be eaten uncooked	< 1/L	<1000/100 ml	A series of stabilization ponds or equiv. treatment
U.S. EPA	Irrigation of pasture for milking animals, fodder, fiber and seed crops and landscape improvement	No standard recommended	200/100 ml[c]	Secondary treatment followed by disinfection
CA	Irrigation of pasture for milking animals, landscape impoundment	No standard recommended	< 23/100 ml[b]	Secondary treatment followed by disinfection
WHO	Landscape irrigation where there is public access, such as hotels	< 1/L	< 200/100 ml	Secondary treatment followed by disinfection
U.S. EPA	Surface or spray irrigation of any food crop including crops eaten raw	No standard recommended	Not detectable[d]	Secondary treatment followed by filtration (with prior coagulant and/or polymer addition and disinfection)

(continued)

Table 2.3 Comparison of the Microbiological Quality Guidelines and Criteria for Irrigation by the WHO (1989), the U.S. EPA (1992), and the State of California (1978) (Continued)

Institution	Reuse conditions	Intestinal nematodes[a]	Fecal or total coliforms[b]	Wastewater treatment requirements
CA	Spray and surface irrigation of food crops, high exposure landscape irrigation such as parks	No standard recommended	<2.2/100 ml[b]	Secondary treatment followed by filtration and disinfection

[a] Ascaris and Trichuris species and hookworms expressed as the arithmetic mean number of eggs/l during the irrigation period.

[b] The California Wastewater Reclamation Criteria are expressed as the median number of total coliforms per 100 ml, as determined from the bacteriological results of the last 7 days for which analyses have been completed.

[c] The number of faecal coliforms should not exceed 800/100 ml in any sample.

[d] The number of faecal coliforms should not exceed 14/100 ml in any sample.

Source: Asano and Levine, 1996.

buildings, and car washing. The piping for effluent use in public gardens must be defined by a purple color. Effluent reuse in the industry is mainly for cooling, cement industry, and fireworks systems. Effluent distributed for aquifer recharge should not pose any risk to the ground water quality or the soil filtering layers.

Besides the national regulations, in the Mediterranean area, no general criteria exist that can be used as reference for all countries. In a workshop held in Crete, Greece (September 25, 2002) wastewater reuse criteria for the Mediterranean region were proposed by the MED-POL working group (A. Bahri, 1999 F. Brissaud et al., 2001). These criteria (summarized in Table 2.4), the "Recycling and Reuse Criteria Proposed for Mediterranean Region," introduce five categories of reuse, providing for each of them the quality requisites and the required treatment process.

2.4 Proposed criteria and guidelines for marginal water treatment and reuse

Criteria and guidelines adopted in several Mediterranean and non-Mediterranean countries (including Ontario, Hawaii, Indiana, Mexico) were collected, compared, and synthesized in order to obtain a set of guidelines and recommendations to be used for a safe and efficient use of marginal water. The attention was mainly focused on the reuse of marginal water for irrigation and aquifer recharge because of their wider application and their greater relevance in

Table 2.4 Recycling and Reuse Criteria Proposed for Mediterranean Region

Category	Type of reuse	Quality criteria	Treatment
I	Urban and residential, landscape and recreational impoundments; also toilet flushing	< 0.1 nematode eggs/ l; 200 ufc FC/100 ml; 20 mg SS/l	Secondary or equivalent + filtration + disinfection
II	Unrestricted irrigation, landscape impoundments (contact with water not allowed), agriculture and industrial reuse	< 0.1 nematode eggs/l; 1000 ufc FC/100 ml; 35 mg SS/l	Secondary or equivalent + storage + maturation or Secondary + filtration or equivalent + disinfection
III	Restricted agricultural irrigation, landscape irrigation with no public access	< 1 nematode eggs/l; FC no standard required; 35 mg SS/l or 150 mg SS/l if coming from lagooning	Secondary or equivalent + filtration + disinfection
IV	Irrigation with application methods providing high degree of protection	< 0.1 nematode eggs/l; 200 ufc FC/100 ml; 20 mg SS/l	Secondary or equivalent + a few days storage + setback distances
V	Ground water recharge	Corresponds, as indicated, to ground water	Site-specific criteria are needed, although for surface spreading primary treatment is required as a minimum For direct injection potable water quality is required An additional condition is issued, indicating that some changes can be accepted depending on the use of water

Source: Bahri, A. and Brissaud, F., 2002.

terms of utilized water volumes. Indeed, industrial and urban reuse as well as marginal water reuse for wetlands creation accounts only for a small percentage of all consumptive use of marginal waters as it is in turn confirmed by the scarcity of related literature. Guidelines and criteria were also prepared for wastewater reservoirs management and design.

2.4.1 Guidelines for the reuse of wastewater in irrigation

Properly planned reuse of municipal and industrial wastewater can alleviate surface water pollution problems and save valuable water resources. The availability of this additional water near population centers can increase the choice of crops that farmers can grow. The nitrogen and phosphorus content in sewage might reduce or eliminate the requirements for commercial fertilizers.

To establish criteria that are valid for a certain region, different local issues and variables should be evaluated, including: availability of primary resources (rain water, surface water, and ground water), water demand of the various productive sectors, type of typical cultivation, nature of the soils, climate, irrigation methods, cultivation techniques, epidemiological conditions and health education of the exposed groups, type and quality of the raw wastewater, efficiency of wastewater treatment plants used, impact of the discharge of the wastewater in surface water bodies, and the cost of construction and operation of treatment plants. It is also extremely important and critical to assess the approval level of interest groups concerning the use of wastewater in agriculture and the effective possibility of marketing the treated wastewater or, in other words, the degree of acceptance by market operators and consumers.

2.4.1.1 Health protection issues

Whenever wastewater effluents are used, health protection measures must be applied. In the past, it was widely accepted that wastewater treatment with some restrictions on crop types would provide enough health protection when using wastewater in agriculture. A well-known study by WHO (1989) indicated that effective health protection can be achieved only by the integration of various control mechanisms, which include wastewater treatment, crop restrictions, control of wastewater application, and human exposure controls.

The main purpose of any health control measure is to protect the people from any direct exposure to pathogens in the wastewater and prevent the spread of diseases. The most vulnerable groups of people when wastewater is used in agriculture include the following:

- Agricultural workers and their families
- Crop handlers
- Consumers of farm products (crops, meat, and milk)
- Those who live nearby the wastewater farm areas

2.4.1.2 Health protection measures

The following different health protection measures should be applied for each group of people:

- Field workers and crop handlers must wear protective clothes and be provided with immunization against selected infections

- Special care should be taken to prevent any accidental use of reused water for domestic purposes, and any sprinklers should not be within 100 m of residential areas or public roads
- Risks to consumers could be reduced by cooking the farm products and maintaining a high standards of food hygiene
- Reliable disinfection to reduce the number of bacteria and other pathogens is another approach. In addition to testing for fecal coliform bacteria, quick tests for chlorine residual can increase confidence in the disinfection system
- Restricting irrigation to times of the day or year when people are not present
- Irrigation only at night or when a facility is closed and establishing buffer areas between the irrigation site and the edge of the field
- Fencing or posting signs to help define the buffer area

2.4.1.3 Nitrogen yield evaluation: Issues and recommendations

All plants use nitrogen (N) to sustain themselves and grow. To encourage plant growth, farmers apply manure or fertilizer to supply the necessary amounts of nitrogen. The amount of nitrogen needed to reach a desired crop yield varies with the crop grown. All of the nitrate and ammonia in the wastewater is available for plant uptake, and any excess can leach into ground water. Organic nitrogen in the wastewater becomes a part of the soil organic matter and is mineralized at a rate of less than 5% per year. It is appropriate to develop a nitrogen balance for the irrigation site to ensure that ground water contamination will not occur.

2.4.1.4 Wastewater reuse system monitoring:
Issues and recommendations

The objective of a wastewater reuse irrigation site monitoring program is to provide for early detection of problems. In most cases, simple adjustments can be made to the operation to avoid polluting ground or surface water.

As a minimum, monitoring should occur at four spots in the system:

1. The treatment plant effluent
2. Storage
3. Irrigation system
4. Soil (and in some cases the vegetation and ground water)

The treatment plant effluent should be monitored to ensure that minimum treatment levels are achieved before it is discharged to the storage facility. The effluent should be monitored for BOD_5, total coliform bacteria, and helminths. Treatment systems using chlorine for disinfection may choose to monitor chlorine residual as an early warning for problems in the disinfection system. Total metal analysis is necessary for treatment plants receiving industrial wastewater. Wastewater flow should also be monitored.

The storage system requires almost the same monitoring of treatment as an effluent one. An additional weekly record of storage volume will help in managing the system to avoid future problems.

The soil within the irrigation site is one of the indicators of all the material being applied. One benchmark site per 5 h can be established and a soil sample can be collected before irrigation each year at the beginning of the application season. For systems over 15,000 m^3/d, samples should be collected twice a year.

By testing a sample of soil from the same spot each year any possible accumulation of minerals and metals can be monitored. This will act as an early warning for possible surface or ground water contamination. If levels begin to get high, simple adjustments can be made in irrigation scheduling to avoid problems.

The vegetation is a biological indicator of all of the material being applied. Both information on yield and plant tissue nutrient levels can act as an early warning system for problems. Plant tissue samples can also be analyzed to reveal nutrient imbalances and the need to add soil amendments such as lime, potassium, or phosphorus.

Ground water should be monitored up-gradient and down-gradient of large irrigation systems. Monitoring wells should be sampled at the beginning and end of the irrigation season for indicators of wastewater contamination.

Monitoring programs for systems greater than 15,000 m^3/d would be similar, but need to be developed individually to meet local conditions and wastewater characteristics. Although much of the monitoring occurs during the irrigation period, some monitoring must continue year-round. Records of wastewater flow and storage volumes, for example, need to be recorded throughout the year. Depending on the pretreatment system used, the effluent may also need to be monitored throughout the year.

2.4.2   Guidelines for the reuse of marginal water for ground water recharge

Artificial recharge can be an interesting option in an integrated strategy to optimize total water resource management. With adequate pretreatment, soil-aquifer treatment, and posttreatment as appropriate for the source and site, impaired-quality water can be used as a source for artificial recharge of ground water aquifers.

Particularly artificial recharge using source waters of impaired quality is a sound option where recharge is intended to control saltwater intrusion, reduce land subsidence, maintain stream base flows, or similar in-ground functions. It is particularly well suited for nonpotable purposes, such as landscape irrigation, because health risks are minimal, and public acceptance is high. Where the recharged water is to be used for potable purposes, the health risks and uncertainties are greater. Although the development of potable supplies has been guided by the principle that water supply should be taken from the most desirable source feasible, indirect potable reuse occurs wherever treated wastewater is first discharged into surface or under-ground water bodies and then withdrawn (downstream or down-gradient) for potable purposes.

These practices should normally be avoided or closely verified and monitored. However, when higher-quality, economically feasible sources are unavailable or insufficient, artificially recharged ground water may be an alternative for potable use.

2.4.2.1 Aquifer characterization: Issues and recommendations

A coordinated, long-term research program should be implemented to support sustainable management of the aquifer and evaluate the real impact of marginal water use for its recharge. The program should emphasize continuity among studies and should be directed by an advisory board with technical representatives from all affected parties having jurisdiction within the area. This program should involve all the institutions that regulate water in the basin, thus bringing different perspectives to the table including environmental, developmental, health, cultural, and scientific interests.

The development of appropriate rates of ground water recharge and withdrawal should be based on social factors, the economics of water resource development and distribution, the influence of conservation and demand-management measures, and public policy.

As a general recommendation, a comprehensive ground water monitoring and protection program should be implemented. Specifically, the long-term study should examine the aquifer-related characteristics of the basin including:

- Identification and mapping of vulnerable areas in the examine aquifer
- Types of human settlements that occur
- Location of active production wells
- Location of abandoned wells
- Type of sewer services provided
- Industries in the area
- Extent of industrial and domestic wastewater treatment employed
- Identification of other activities that contribute to ground water contamination

The long-term study should also examine the thickness, extent, and depth of the aquifers as well as estimate the porosity, permeability, storability, and hydraulic conductivity of the aquifers. Other important components include:

- Changes in water quality with depth, geographic location, and relation to producing well fields
- Degree of connectivity between the various zones within the aquifers and in the recharge zones
- Extent and location of faults or other compartmentalizing factors within the aquifers important for optimizing well placement

- Physical, chemical, and biological characterization of the aquifers
- Identification of the critical water levels below, with which continued pumping would no longer be efficient, is required to predict the behavior of the aquifer

After characterizing the aquifer with some level of confidence, an interagency and interdisciplinary panel should determine an optimum yield for the aquifer on the basis of an evaluation of multiple objectives. It may be useful to engage in this analysis the same advisory board that would be directing the long-term ground water research program for the basin. What is optimal for the aquifer will depend, at a minimum, upon a number of interrelated factors:

- Consideration of the economic dependence of the region on the ground water resource
- Consideration of deteriorating water quality with increasing recharge rate of marginal waters
- Consideration of deteriorating water quality with increasing aquifer depth
- Current impacts of other point source and nonpoint source pollution
- Availability and actual marginal cost of obtaining and distributing other new sources of water under normal and drought conditions
- Influence and potential of programs for water pricing and metering, water conservation, water reuse, and ground water recharge
- Impact of water use on other environmental interests under normal and drought conditions
- Best calculations available as to the potential long-term life of the aquifer at the various rates of recharge and pumping based on the considerations above

The capabilities for water quality data collection, information storage, and reporting of monitoring results should also be improved. Current and reliable information should be available to the general public as well as government and research institutions. The information should be at a suitable level of detail to identify what parameters may be out of compliance in specific areas of the distribution system and its significance to public health.

2.4.2.2 *Recharge techniques: Issues and recommendations*

Once recharged marginal water has been deemed feasible as part of an integrated approach to regional water supply planning, the method of recharge chosen should be based on hydrogeologic and sanitary conditions and the specific benefits sought from the recharge. Surface spreading would be preferred as an aquifer recharge method, as it offers the greatest engineering and operational advantages, including:

- Surface methods can accommodate waters of poorer quality and are simpler to design and operate than recharge wells, although certain conditions may require use of wells
- Surface spreading requires large amounts of land with permeable soil, it may not be feasible in densely populated areas or where suitable land is expensive or unavailable
- Injection wells require high-quality source water to avoid clogging problems and also because aquifers alone do not provide the same degree of treatment as soil-aquifer systems

Although there are indications of some water quality improvements within aquifers, considerable pretreatment is necessary because of the source water's impaired quality.

Artificial recharge of ground water using source waters of impaired quality to augment water supplies should be considered primarily for non-potable uses, since it might help to reduce the demand of limited freshwater sources at minimal health risk; thus, it is widely practiced and accepted.

Careful preproject study and planning, especially where potable reuse is considered, is required. Specifically, artificial recharge of ground water with waters of impaired quality should be used to augment water supplies for potable uses only when better-quality sources are not available, subject to thorough consideration of health effects and depending on economic and practical considerations.

Treated municipal wastewater, stormwater runoff, and irrigation return flows are the main types of impaired quality waters potentially available for ground water recharge. The following consideration may apply:

- Treated municipal wastewater is usually the most consistent in terms of quality and availability
- Stormwater runoff from residential areas generally is of acceptable quality for most recharge operations, but at some times and places it may be heavily contaminated, and its availability is variable and unpredictable
- Irrigation return flow exhibits wide variations in quality and is some-times seriously contaminated and thus usually is not a desirable source of water for recharge
- Based on current information, treated municipal wastewater intend-ed as a source for artificial recharge should receive at least secondary treatment
- Municipal wastewater that has received only primary treatment may be adequate for the recharge of nonpotable ground water in certain areas, but use of primary effluent should not be considered without implementation of a site-specific demonstration study

Certain impaired quality waters, such as irrigation return flow, storm-water runoff from industrial areas, and industrial wastewater, generally should not be considered as suitable sources for artificial recharge. Exceptions might be identified, but only after careful characterization of source water quality on a case-by-case basis. Other types of stormwater runoff to avoid include most dry weather storm drainage flow, salt-laden snowmelt flow, and flow originating from certain commercial facilities, such as vehicle service areas. Construction site runoff also should be avoided to prevent clogging of recharge facilities with eroded soil and other debris.

2.4.2.3　*Human health protection: Issues and recommendations*

The main concern regarding artificial recharge by using waters of impaired quality for potable purposes is the protection of human health. There are uncertainties in identifying potentially toxic constituents and pathogenic agents, and thus potable reuse should be considered only when better quality sources are unavailable.

Care must be paid to the possible disinfectant by-products in treated marginal waters, and specifically:

- Disinfection of treated municipal wastewater prior to recharge should be managed in order to minimize the formation of disinfectant by-products
- Alternatives to chlorination include disinfection with ultraviolet radiation and the use of other chemical disinfectants

However, additional research should be undertaken on pathogen removal, formation of disinfectant by-products, and removal of disinfectant by-products before alternative disinfectants can be classified conclusively superior to chlorine.

In addition, continuous monitoring of ground water quality characteristic is required. Specifically recovered water must be monitored carefully to ensure that pathogenic microorganisms and toxic chemicals do not occur at concentrations that might exceed drinking water standards or other water quality parameters established specifically for reclaimed water that consider the nature of the source water.

Further research in health risk assessment is necessary due to the significant uncertainties associated with the transport and fate of viruses in recharged aquifers. These uncertainties make it difficult to determine the levels of risk of any infectious agents still contained in the disinfected wastewater.

2.4.3　*Guidelines for marginal water urban reuse*

Implementing a new reclaimed water distribution system in developed urban areas can be too expensive. In some specific cases, however, the benefits of the savings of drinkable water can justify the costs. For example, the system can have sustainable costs if it eliminates or limits the need to use resources located at significant distances. In new developments, instead, the dual system installation can constitute an advantageous choice.

When planning an urban reclaimed water distribution system, one of the most important considerations concerns the reliability of the service to safeguard public health. The protection of public health can be obtained by treating the wastewater effluent to guarantee a reduction of the concentration of pathogen bacteria, parasitic, enteric virus, and chemical constituents that can be dangerous for human health. In this case the level of treatment also depends on the specific use. However, since for many urban water reuse options the contact with humans is not excluded, the minimum required treatment level must be tertiary (for example, filtration), followed by an appropriate disinfectant (chlorination, ozone disinfection, UV radiations).

The following strategies/measures should be incorporated in the design of any dual distribution system:

- Ensure that the reclaimed water delivered to the customer meets the water quality requirements for the intended uses
- Prevent improper operation of the system
- Prevent cross connections with potable water lines
- Equipment associated with reclaimed water systems must be clearly marked, and nonpotable pipelines must be characterized by specific coloration to avoid cross connections

2.4.4 Guidelines for marginal water industrial reuse

Generally, industrial water users require significant amounts of quality water for a large variety of uses. As a minimum, secondary reclaimed water is recommended as a basic substitute to potable water offered to a target industry. Beyond that level, each specific industrial use may impose its own particular set of water quality requirements. The quality requirement for each use depends on the industry's specific water demand characteristics, type of process in use, cycles of in-plant reuse, and type of product manufactured.

When marginal waters are used for cooling, pathogenic microorganisms present potential hazards to workers and the public in the vicinity from aerosols and windblown spray, especially if not well-disinfected wastewater is used. In practice, however, biocides are usually added to all cooling water onsite to prevent slimes and otherwise inhibit microbiological activity, which has the secondary effect of eliminating or greatly diminishing the potential health hazard associated with aerosol or windblown spray. *Legionella pneumophila*, the bacterial agent that causes Legionnaire's disease, is known to proliferate in air conditioning cooling water systems under certain conditions. All cooling water systems should be operated and maintained to reduce the *Legionella* threat, regardless of the origin of marginal water source.

2.5 Cost analysis for marginal water treatment

The evaluation of capital and operation and maintenance costs is a fundamental phase in the planning of marginal water reuse project. Unfortunately, marginal water treatment costs are not well documented. In this work, an attempt to obtain the unit treatment costs for marginal water (mainly waste-water), using national (Italian) and international published data, has been carried out.

The costs for the different treatment processes that might be required for the various reuse alternatives can be broken down into two main components: the initial investment cost and the operation and maintenance costs. The initial investment costs can be further broken down into the following items: land acquisition costs; civil works costs; and electromechanical equipment costs. On the other hand, operation and maintenance (O&M) costs can be broken down into:

- Manpower wages and salaries
- Power consumption
- Sludge treatment and disposal
- Ordinary and extraordinary maintenance
- Chemicals (chlorine or disinfectants, metal salts, polyelectrolytes)

In order to provide actual costs, taking into account the variation with the size of the treatment plant, the following typologies of secondary treatment are considered:

- Activated sludge with aerobic sludge stabilization (for potentiality < 50,000 equivalent inhabitants);
- Activated sludge with anaerobic sludge stabilization (for potentiality 50,000 equivalent inhabitants)

Chlorination is considered the best disinfectant process for both treatment schemes. These two conventional schemes were used to calculate the treatment

Table 2.5 Equation of Unit Treatment Cost Curves

Treatment alternatives	X < 30.000 E.I.	X = 30.000 E.I.
Primary treatment	$Y = 0.317 - 9 \times 10^{-6} \cdot X$	$Y = 0.132 - 5 \times 10^{-7} \cdot X$
Secondary treatment	$Y = 0.474 - 7 \times 10^{-6} \cdot X$	$Y = 0.309 - 4 \times 10^{-7} \cdot X$
Filtration	$Y = 0.507 - 7 \times 10^{-6} \cdot X$	$Y = 0.342 - 4 \times 10^{-7} \cdot X$
Nitrification/ denitrification + filtration	$Y = 0.559 - 8 \times 10^{-6} \cdot X$	$Y = 0.369 - 5 \times 10^{-7} \cdot X$
Nitrif./denitrification + phosph. removal + filtration	$Y = 0.602 - 8 \times 10^{-6} \cdot X$	$Y = 0.393 - 5 \times 10^{-7} \cdot X$
Coagulation-flocculation	$Y = 0.939 - 2 \times 10^{-5} \cdot X$	$Y = 0.471 - 5 \times 10^{-7} \cdot X$
Carbon adsorption	$Y = 1.132 - 1 \times 10^{-5} \cdot X$	$Y = 0.730 - 5 \times 10^{-7} \cdot X$
Reverse osmosis	$Y = 1.503 - 2 \times 10^{-5} \cdot X$	$Y = 0.907 - 5 \times 10^{-7} \cdot X$

Note: Y indicates the unit costs in €/m³; X indicates the number of equivalent inhabitants (E.I.);

For $x < 1000$ E.I., a constant cost is assumed equal to that obtained for 1000 E.I.;

For $x > 200,000$ E.I., a constant cost is assumed equal to that obtained for 200,000 E.I.

costs for the other treatment alternatives, including advanced treatments such as filtration, coagulation-flocculation, activated carbon adsorption, and reverse osmosis.

The collected data on treatment costs have been used to develop the unit treatment cost curves as a function of treatment plant size. The unit treatment costs, expressed as €/m³ of treated water, include investment and management costs. The investment costs have been amortized by an interest rate of 5% and assuming a life of the treatment plant of 30 years for the civil works and a replacement every 10 years of mechanical parts. As an example, the cost curves for the different treatment alternatives are given in Table 2.5 assuming a daily per-capita discharge of 300 L/E.I.-d.

2.6   Development of a web-based information system for wastewater treatment and reuse

Among marginal waters, wastewater is rather available and it can be reused advantageously in several applications after appropriate treatment. As highlighted in the previous paragraphs, the level of treatment required to guarantee the necessary protection of public health depends on specific reuse options, and it should be established by apposite regulations, criteria, and guidelines. Considering the numerous issues and the nonhomogenous criteria and guidelines, it was considered useful, as part of this project, to develop a web-based information system (WBIS) summarizing and organizing the main information indispensable for wastewater reuse projects. This tool called E-Wa-TRO (evaluation of wastewater treatment and reuse options) was developed as a website and is targeted to water management authorities and private users with the aim to provide, for each specific wastewater reuse alternative, the fundamental information for a feasibility study, such as existing regulations and guidelines, best available technologies of treatment, technical, and hygienic issues, use-recommendations and expected capital, operation, and maintenance costs (Sipala et al., 2003).

2.6.1   Development and implementation

One of the design strategies was to implement a system that, interacting with the user, could include a great variety of hypothetical alternatives, being able to supply, for the different categories of reuse, the necessary information to identify the most convenient application. It was therefore decided to include in the WBIS the following information:

- Existing standards for wastewater reuse in Mediterranean and non-Mediterranean countries
- Water quality requisites for each specific reuse alternative
- Appropriate treatment technology to achieve the required water quality requisites
- Treatment costs, including capital, operation, and maintenance costs

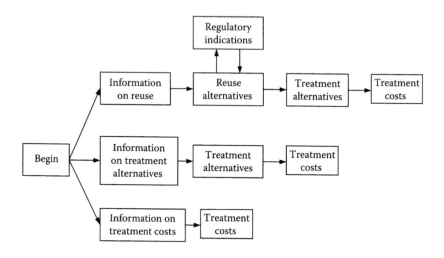

Figure 2.1 Layout of the web-based information system E-Wa-TRO (Sipala et al., 2003).

Figure 2.1 shows a simplified flowchart of E-Wa-TRO. Depending on the goal of the user, the WBIS will permit the user to acquire the necessary information to evaluate the feasibility of a reuse project, or to obtain a complete picture of the different alternatives available in wastewater treatment technology, along with the related treatment costs (these were evaluated on the basis of data and calculations presented in the previous paragraphs of this chapter).

Regarding specific reuse alternatives, the E-Wa-TRO can supply the necessary indications, based on specific inputs of user as outlined in Figure 2.2. For example, with regard to agricultural reuse, the user can choose the irrigation technique and crop type, obtaining the qualitative characteristics for reusable water and the minimum level of treatment. Information on treatment costs and regulatory indications are also available.

E-Wa-TRO was implemented as a website by means of a series of separated files (total 65), formatted using the HTML language, connected through hypertext links. Such implementation gives many advantages, including:

- Simple consultation even for nonexpert users
- Wide diffusion to a vast public
- Easiness of modifications and integrations

Each file in the site contains general information and several links allowing the user to choose among the various alternatives. The E-WA-TRO home page, accessible through http://www.dica.unict.it/users/fvaglias/EWATRO/ is shown in the Figure 2.3.

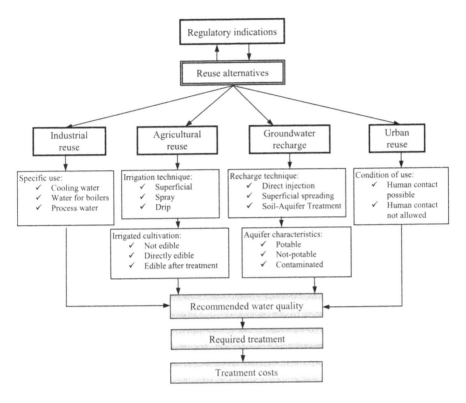

Figure 2.2 Website structure (empty boxes represent interaction phases with the user; gray boxes represent information acquisition by the user).

2.6.2 E-Wa-TRO application

In order to illustrate the potentialities of the E-Wa-TRO, two reuse scenarios were hypothesized, and an application of the web-based tool was carried out. A farm for the production of edible horticulture (such as tomatoes, cucumbers, etc.) and a textile industry were considered to evaluate the reuse of water to irrigate the crops of the farm or to supply process water from a wastewater treatment plant for the textile industry (supposed to be in Sicily, Italy), with a capacity of 50,000 EI.

The E-Wa-TRO is first used to identify the applicable water reuse regulations. From the index on the left, the page with regulatory overview can be chosen. Wastewaters reuse standards of Mediterranean and non-Mediterranean countries are available on this page.

Choosing the Mediterranean countries link, a set of national standards for wastewater reuse can be found. In this example, since the treatment plant is localized in Sicily, the page with the national Italian, and Sicilian regulations can be accessed and the pertinent regulatory indications can be acquired.

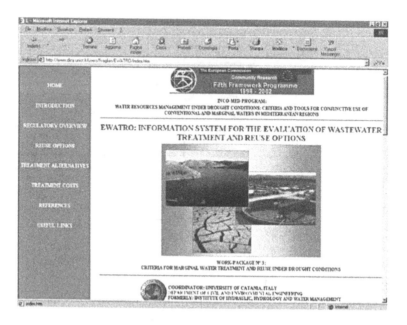

Figure 2.3 Home page of the web-based information system E-Wa-TRO (Sipala et al., 2003).

After acquiring the regulatory information, the E-WA-TRO provides detailed information on the water quality requirements, treatment alternatives, and resulting costs. In the case of the farm, by selecting the various agricultural reuse alternatives, it is possible to access the page related to the irrigation technologies. On the basis of, the regulatory limitations, localized irrigation has to be chosen. Based on the specific cultivation to be irrigated (crops to be consumed raw), the appropriate page gives the required qualitative characteristics for the water to be reused and the minimum level of treatment (in this case, a secondary treatment followed by a filtration phase). By activating the link to the costs of treatment and giving as input data, the size of the plant (e.g., 50,000 EI) and the water supply per capita-per day (e.g., 250 l/EI-d, which is a common value in Sicily), it can obtain the unit cost of treatment equal to 0.36 € / m^3 (Figure 2.4).

Similarly, for the case of the textile industry, industrial reuse can be selected among the reuse alternatives, therefore accessing the page where several applications of wastewaters in this sector (cooling water, water for boilers, process water) are described. After selecting the process water alternative, the page related to the qualitative characteristics of the water to be used in the textile industry and the advised form of treatment is accessed, where an advanced treatment with reverse osmosis is suggested. Using the same inputs utilized for the previous scenario, a unit cost of treatment equal to 0.89 € / m^3 (figure not shown) is obtained.

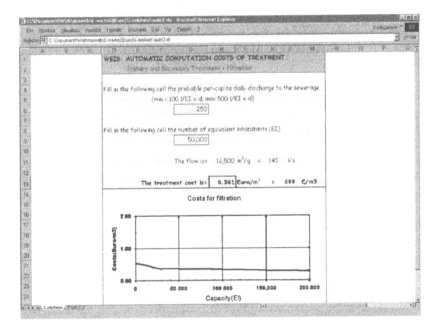

Figure 2.4 Page showing the results from the automatic computation of cost for filtration (Sipala et al., 2003).

The information provided by the E-Wa-TRO could represent the basis for supplementary evaluations that will be carried out, taking into account the local conditions (other water resources availability, rates to pay for these resources, etc.). Therefore, the E-Wa-TRO allows the user to orientate him- or herself among the various reuse alternatives, the treatment technology available, and the mandatory regulation. Moreover, the specifically calculated treatment costs represent a good basis for evaluating the most convenient reuse alternative.

The E-Wa-TRO could be upgraded in the future to include supplementary information such as the costs of hydraulic infrastructures necessary for the storage and transport of the treated wastewater to the reuse sites.

2.7   Conclusion

In this work potential applications and limits for the use under drought conditions of marginal waters were investigated, and appropriate reuse criteria and guidelines were identified. Attention was focused on the possible uses of the marginal resources and the related issues. Specifically, agricultural irrigation, ground water recharge, and industrial reuse were examined in details outlining the requisites for the water to be utilized. Also some information on urban reuse and natural and manmade wetlands were provided.

An analysis of treatments costs was examined, providing both capital and management costs for the treatment processes that can be used to achieve the quality requisites necessary for the safe and advantageous reuse of marginal water.

The main purpose of this chapter was the definition of criteria and guidelines for marginal water treatment and reuse under drought conditions. This task was accomplished by first comparing the criteria and guidelines existing both in non-Mediterranean and Mediterranean countries as well as those provided by international organizations in order to establish criteria valid for different geographical regions and local realities. A synopsis of the collected information was then prepared. Specific criteria were provided for marginal water reuse for irrigation and aquifer recharge, whereas more general ones were provided for industrial and urban reuse and marginal water reuse for wetlands creation.

All the information collected and the results obtained in terms of treatment cost analysis, criteria, and guidelines for wastewater treatment and reuse were consolidated into a web-based information system for the evaluation of wastewater treatment and reuse options, called E-Wa-TRO. This tool can be used by water management authorities and private users to get, in a simple and immediate way and for each specific reuse alternative, the most important information for a feasibility study, such as the existing normative and guidelines, the best available treatment technologies, the technical and hygienic issues, and the expected capital, operation, and maintenance costs.

On the basis of the information collected and on the different reuse scenarios simulated with E-Wa-TRO, it can be concluded that marginal water treatment and reuse may represent a viable alternative resource within integrated water management under drought conditions. However, cost of treated marginal waters may prove prohibitive especially in those countries where strict regulations for reuse are enforced. As a result, it appears necessary to identify appropriate marginal waters and the necessary treatment as a function of the reuse application. This approach could minimize the cost associated with reuse while maintaining at acceptable low levels the risk for public health.

2.8   Acknowledgment

This project was partially funded by the EU through the program INCO-MED of the V Framework Program, contract ICA3-CT-1999-00014 (project: Water Resources Management Under Drought Conditions: Criteria and Tools for Conjunctive Use of Conventional and Marginal Waters in Mediterranean Regions), and by the Italian Ministry of Instruction, University, and Scientific Research (MIUR), through the program Research Projects of National Interest, Fiscal year 2001 (Project: Integrated Water Cycle: Measures to Ameliorate the Quality of Water Resources).

References

Anderson, J., Adin, A., Crook, J., et al. (2001). Climbing the ladder: a step by step approach to international guidelines for water recycling. *Water Sci. Technol.*, 43(10).

Asano, T., and Levine, A. D. (1996). Wastewater reclamation, recycling, and reuse: Past, present, and future. *Water Sci. Technol.*, 33(10), 1–14.

Bahri, A. (1999). Agricultural reuse of wastewater and global water management. *Water Sci. Technol.*, 40(4–5), 339–346.

Bahri, A. and Brissaud, F. (2002). Issue and Criteria for Water Recycling. Workshop on Water Recycling and Reuse Practice in Mediterranean Countries, Crete, Greece, September 25, 2002.

Food and Agricultural Organization of the United Nations (FAO). (1992). *Wastewater treatment and use in agriculture.* M. B. Pescod. FAO Irrigation and Drainage Paper 47, FAO, Rome. 125 p.

Hammer, D.A., and Bastian, R.K. (1989). Wetland Ecosystems: Natural Water Purifiers. In *Constructed Wetlands for Wastewater Treatment*, D.A. Hammer (Ed), Lewis Publishers, MI. 2, 5–19.

Lazarova, V., Levine, B., Sack, J., et al. (2001). Role of water reuse for enhancing integrated water management in Europe and Mediterranean countries. *Water Sci. Technol.*, 43(10).

Sipala S., Mancini G., and Vagliasindi F.G.A. (2003). Development of a Web-Based Tool for the Calculation of Costs of Different Wastewater Treatment and Reuse Scenarios. *Water Sci. Technol.*, 3(4).

U.S. Environmental Protection Agency (EPA). (1992). *Manual, guidelines for water reuse.* EPA 625/R-92/004. Washington, DC: Office of Water: Office of Wastewater Enforcement and Compliance; Cincinnati, OH: Office of Research and Development; Office of Technology Transfer and Regulatory Support Center for Environmental Research Information.

World Health Organization (WHO). (1989). Health Guidelines for the Use of Wastewater in Agriculture and Aquaculture, Technical Report Series 778, WHO, Geneve, Switzerland.

chapter three

Strategies for the conjunctive use of surface and groundwater

Andres Sahuquillo
Universidad Politécnica de Valencia, Spain

Contents

3.1 Introduction

Ground water is an important hydrological component of watersheds. Average river flow drainage from aquifers in continental areas is in the order of 30% of total stream flow, which is essential in sustaining stream flow during dry periods, the so-called base flow in permanent rivers. Magnitude of aquifer recharge, the usually big volumes of water stored in them, easiness of their exploitation, and the overall much lower cost of ground water development make their use very attractive.

Wise use of the different and complementary characteristics of surface and subsurface components through conjunctive use of surface and ground water can achieve greater yields, economic, or functional advantages than separate management of both components. One complementary characteristic is the large volume of water stored in aquifers, from tens to hundreds of times their annual recharge. In the same way, volume of aquifer storage provided by a relatively small fluctuation of the piezometric head in unconfined aquifers considerably exceeds the available or economically feasible surface storage. That allows the use of water in storage during dry seasons as well as the use of the subsurface space for storing surface or subsurface water. The existence of aquifers over ample areas of a basin adds to the benefits of water storage those of distribution and conveyance. Moreover, long-term storage in and passage through a ground water aquifer generally improves water quality by filtering out pathogenic microbes and many, although by no means all, other contaminants.

Many uses are common to both surface and ground water (irrigation, municipal and industrial uses, and joint ecological benefits such as wetland maintenance). In fact ground water has traditionally been used worldwide to create a supply for times of shortage, being in some way a kind of conjunctive use. In those cases ease of implementation and efficiency is obtained with insignificant investments that are in most cases peerless as compared with those usually required for implementing structural alternatives to attain similar objectives. Similarly important advantages can be obtained with more comprehensive conjunctive use of ground water and surface water. Ground water can produce other unique environmental benefits related to base flow and riparian habitat preservation. In addition, ground and surface water are hydraulically connected, so the contamination of one can migrate to the other. In relatively complex systems, these advantages do not appear so evident simply because in very few cases a comparison of different alternatives, including conjunctive use, has been made using simple tools.

The use of ground water can serve, and in some cases has been used purposely, to defer the construction of costly surface water projects even at the expense of temporary overdrafting the aquifer. In others cases, high volumes of water stored in the aquifers had been allowed, through unplanned overdraft, to sustain primary economic activities, which resulted in further economic growth.

Another unquestionable argument in favor of the joint consideration of ground and surface water is the fact that to a greater or lesser extent they are hydraulically connected. Infrastructures that use surface water and ground water affect each other as well as other components of the hydrologic cycle. Ground water recharge can be augmented by storing water in leaky surface reservoirs, by transporting water in unlined canals, or by return flow from irrigation. On the other hand, recharge to underlying aquifers from losing streams can decrease as a result of water being diverted upstream. Due to the changes produced in the sequences of river flows, surface storage can increase or decrease the recharge in downstream aquifers located beneath losing reaches of the river channel. Ground water pumping can cause depletion of surface or spring flow and can produce other externalities such as land sub-sidence or destruction of riparian habitats and wetlands. These effects can produce environmental, legal, and economic problems that must be addressed. In most of these scenarios conjunctive use is suitable for bringing out the positive effects and playing down the negative ones (NRC, 1997).

Excessive return flow irrigation and canal losses in arid areas have produced extensive drainage problems and an increase in salinity in many areas of the world. Conjunctive use can help to solve or attenuate these problems with the additional advantage of increasing the safe yield of the system with the use of the augmented ground water recharge from canal losses and return infiltration.

The strongest argument in favor of conjunctive use is that aquifers pro-vide alternatives, not only for augmenting the number of components but above all, for increasing their functionality and therefore the probability of being more effective. Although in most areas ground water is hardly con-sidered by managers, it can provide useful solutions to many problems. Likewise conjunctive use can be applied to obtain a better or cheaper solution to existing problems. Its suitability must not be restricted to application in only arid or water scarce areas. On the contrary, if surface and ground water relationship and mutual influence are considered, conjunctive use is advis-able in most areas including where scarcity or pollution problems exist.

Aquifers can be a source of water as well as perform complementary functions of storage, water distribution, and treatment, which are classical components of a surface system. In aquifers, the water distribution role is directly related to the storage function. A conjunctive use system of both surface and subsurface components dynamically conceived, expanded, and operated to keep up with water demand, and hydrologic variability can pro-vide economic, functional, and environmental advantages. To quantify the potential benefits, many alternatives have to be analyzed by means of more efficient, simple, and easy to understand comprehensive models. Water quality and contamination have only been indirectly or qualitatively considered in conjunctive use analysis. Only in some cases have total water salinity or gradients restriction used as surrogate parameters been explicitly modeled.

In recent years in most developed and developing countries structural solutions are being questioned, and there is a growing trend in favor of better management of existing elements instead of large investments in new constructions. In many countries the time of constructing new dams has passed. The most favorable and less controversial sites have been already built to keep pace with a higher environmental conscience. Additionally large-scale hydraulic constructions can cause legal, economic, and social problems. In many cases, big investments can create grave financial problems to some developing countries. Recently conjunctive use alternatives are being considered prior to enlarging existing water resources.

From another perspective, one would see the conjunctive use of surface water and ground water as being a mechanism through which the use of available water resources is optimized, and the benefits of doing so are greater than if both sources were managed in an uncoordinated manner. It has to be clear that uncoordinated simultaneous use of surface and ground waters should not be considered as conjunctive use (although this is a frequent misconception). Conjunctive use at least involves decisions on when, where, and in which amount to use each one of the sources of water. It has been demonstrated (Sahuquillo and Lluria, 2003) that such a coordinated use of both resources may help to solve, or at least attenuate, water quality and water quantity problems. Most often, conjunctive use can prove to be a cheaper solution than sole dependence on either surface water or ground water.

Among the advantages of the conjunctive use of available water resources are the economic, operational, and strategic benefits, or improvements, a society might obtain when optimizing both resources. Although not very obvious at the start of a project, the economic advantages become clear when new investments for water supply sources (construction of large dams) decrease and the operational costs of integrated systems are lowered. The operational advantages include the increase of available water resources for water supply without necessarily increasing the storage in the basins. Furthermore, some problems, due to overexploitation of either surface or ground water resources, may be solved or at least mitigated, such as the drainage and salinization of soils in irrigated lands in arid and semiarid regions, land subsidence due to excessive pumping, and so forth.

From our experience with many cases analyzed during the past 20 years in Spain and other countries, when there is a significant ground water component somewhere in the system, some advantages are always achieved. Depending on each case, when ground water resources or the surface extension of the aquifers in the basin is important, advantages usually became decisive. The purpose of this chapter is to discuss how conjunctive use can increase the water availability in the developing world, what types of conjunctive use schemes are more promising, and also to present tools and models developed in the Department of Hydraulic and Environmental Engineering of Polytechnic University of Valencia to analyze in an integrated way the basin performance for conjunctive use cases, emphasizing their easiness to use, versatility, and rigor.

As with most human activities, the practice of conjunctive use is subject to, and governed by, many political, social, and economic factors. The advantages to be obtained by putting conjunctive use into practice depend on physical factors, but rules and institutions permit or hamper its use. Rules governing water use, such as laws defining water rights, are critical. Water rights affect incentives for involvement in conjunctive management. We will not discuss the legal and institutional factors that have been addressed elsewhere (Sahuquillo and Lluria, 2003), but it is necessary to keep them in mind.

3.2   Methods of conjunctive use

There are two possibilities for using the storage provided by aquifers. The most intuitive is through artificial recharge. The second is through alternate use of ground water and surface water. In alternate conjunctive use (ACU), target yield is obtained in dry years through increased pumping; when more than average water is available in streams or surface storage, more surface water is used, allowing more ground water to remain in storage. Operating in this way, storage is provided through differences between extremes of the aquifer water levels, these being high at the end of wet periods and low at the end of dry ones. Both possibilities of artificial recharge and alternative use are not exclusive. In fact there are many sites where both are applied although one of them usually predominates.

The rationale behind adopting an approach of conjunctive use of water resources are mainly, although not exclusively, to take advantage of the storage capacity of aquifers, the hydrological interlinkages between ground water and surface water, and the differences in the timing of water circulation between these water bodies. The main basic schemes for conjunctive use include artificial recharge and ACU.

3.2.1   Artificial recharge

The rationale of subsurface storage in artificial recharge is very clear. The usual practices of artificial recharge are through injection wells and infiltration ponds. In arid regions, artificial recharge is an appropriate option, but this practice may also be applied in other areas and for other purposes. Artificial recharge has been used in past times to store surface flows or nonused surplus water that otherwise would be lost. More recently it has been used to improve aquifer management, including reduction of water levels descent, seawater intrusion recovery, and others. In many countries of northern and central Europe aquifers are widely used taking advantage of soil and vadose zone faculty to filtrate and treat polluted recharged surface water. In this chapter that practice is not considered as conjunctive use. The objective of artificial recharge is to stop land subsidence caused by ground water head depletion and others related with sewage water treatment and reclamation or with environmental and contamination problems, which in this chapter is not considered a particular type of conjunctive use. On the contrary, the objective of mixing in the aquifer waters with different chemical composition to dilute chloride, nitrate, or other contaminants is an interesting, although not very commonly used, conjunctive use scheme. It is practiced in Israel where the imported water of the Kinneret lake is more salty than the water in the coastal and calcareous aquifers where water is recharged to be stored. Artificial recharge of surface water with low nitrate content has been proposed in La Plana de Castellón aquifer in Spain in order to lower its high nitrate levels.

In Israel, in a planned way, and spontaneously in Southern California, aquifers were overexploited from the early stages of hydraulic development. Soon scarce local surface water and later imported water were recharged into aquifers. Artificial recharge has been employed in many arid areas in the world, but it is in the above-cited areas where artificial recharge has been used extensively. In further stages, sewage treated effluent has been recharged in some aquifers after having passed advanced treatment. In Southern California in the wells of the hydraulic barriers constructed to protect some coastal aquifers from seawater intrusion, and in Israel treated sewage water from the metropolitan area of Tel Aviv is recharged in sand dunes to be pumped later for accepted uses. In the arid and semiarid regions of the western U.S., the predominant artificial ground water recharge method is direct surface recharge, frequently referred to as water spreading. This consists of direct percolation of the surface water from recharge basins constructed on highly permeable soils to the aquifer. The origin of the recharged water could be from local rivers and their tributaries, from municipal, industrial, and agricultural recycled water, from desalted water, or from an imported water source.

Artificial recharge is usually expensive, both for wells and infiltration ponds. There is in general need of desilting and treating the water to be recharged to avoid clogging, and it is necessary to clean and unclog ponds and wells. After some time the recharge capacity of wells cannot be regenerated to operative flows, and they have to be replaced. Infiltration in losing rivers, ephemeral streams, and alluvial fans can be important in many cases, and there exist possibilities to economically enhance it. The origin of recharged water can be settled, or unsettled, surface runoff, or water stored in reservoirs timely discharged to losing river channels. Unintended aquifer recharge from pervious reservoirs in some Mediterranean karstic areas in Spain became very advantageous, and the possibility of purposely building some has been suggested in several sites.

By far, it is in California where more water is recharged, around 3000 million cubic meters per year. In Spain artificial recharge without any doubt will be used in the near future in more sites to solve some local problems, but it is not expected to solve any major problems. Alternative use schemes, as implemented in many other countries, appear to be more attractive as will be discussed later. Artificial recharge requires adequate technical operation and monitoring and permanent supervision. In less economically and technically developed semiarid regions, the influence of operation and maintenance in final water cost could be high for most irrigation needs.

The method known as aquifer storage recovery (ASR) was first employed in the state of Florida; it is used predominantly for drinking water supply. It consists of the underground storage of treated water during periods of low demand and its recovery for potable water uses during periods of high demand. The recharge operation is carried out with dual-purpose wells that inject the water into the aquifer and also recover it by pumping. This method is well suited for use in areas where direct surface recharge is not applicable (Pyne, 1989). A similar concept is used in the ground water reservoir situated in the Palaeogene sands and chalk aquifers existing beneath the London clay in the Thames river. The aquifer was first exploited in the 18th century. Over the next 200 years the aquifers were heavily pumped. The water level gradually fell, and saline water from the tidal river Thames intruded into the aquifers; but the chalk aquifer is still used in the Lee Valley and is recharged through wells during the winter with treated water from the rivers Thames and Lee. The same temporary storage function of treated potable water is used in Barcelona. Up to 20 million cubic meters per year are recharged by dual-purpose wells, to be stored in the Llobregat Delta aquifer when water tanks of the raw water treatment plant are full (UK Groundwater Forum, 1998; Custodio et al., 1969).

Water banking is a concept in the water management literature that is firmly related to artificial recharge. It can be defined as an operation that stores surface water in aquifers by artificial recharge techniques during wet years or when surface water from importation or recycling is available in surplus quantities and extracts it for use during dry periods or when water demand has increased beyond the forecast annual level. The concept of in-lieu recharge is often considered a type of conjunctive use. We consider that its guiding idea is the same as the alternate use.

3.2.2 *Alternate conjunctive use*

A frequent misconception among hydrologists and water planners is to identify conjunctive use mainly with artificial recharge practices. In most cases ACU is much cheaper and easier to implement than artificial recharge, particularly in developing countries (Rivera et al., 2005). ACU is a simple type of conjunctive use, whereby surface water is used preferentially in wet periods, and ground water is used preferentially in dry periods. However, pure surface water demands, pure ground water demands, and alternate water demands usually coexist. The use of subsurface storage is achieved by differences in storage between the higher levels at the end of several wet years with important ground water recharge and less pumping, and the lower levels at the end of a dry period with less recharge and considerable abstractions from the aquifer. The concept is less intuitive than artificial recharge, but in no way less effective and in most cases much cheaper. ACU is currently applied in coastal aquifers, large interior aquifers, alluvial aquifers, and in the "drought supplemental wells" approach. In less-developed semiarid regions, it could be a better option than artificial recharge because in general it is more economic, has less technical problems, and is more suitable to developing countries. Moreover, in addition to being more costly and complex in operation, artificial recharge needs a clear identification of investors and beneficiaries, and it needs a complex technical and institutional development. These conditions are infrequent in developing countries. Nevertheless, that does not preclude the convenience in many cases of the enhancement of natural recharge or development of methods to lower the cost of artificial recharge.

In ACU ground water is used more often in dry periods, contrary to its decreased use and surface water use augments when there is more surface water available in rivers or stored in surface reservoirs. In that type of conjunctive use a part of the water demand can be supplied by more than one source. As a portion of the water demand is supplied alternatively from different sources according to the situation of each component, whether it is surface or subsurface, the system can satisfy a higher water demand.

Ground water has traditionally been used in many countries to supplement scarce surface water supplies during drought periods, with the improvement in the reliability of the system achieved by using ground water at the right moments being of even greater value than the augmentation of supply. Without augmenting surface storage some conjunctive use schemes utilize that possibility to augment the firm yield. If firm yield requirements increase, during the same critical period in which reservoirs fail to provide the required supply, an increase in ground water pumping during larger periods is needed. Similarly for a fixed firm yield, reliability can be augmented with additional increases of pumping. Water availability as well as ground water in storage can be increased using more surface water during wet years, diminishing during ground water pumping as much as possible, in areas where aquifers are used in dry or not as wet years. In many cases some new connecting element has to be created or enlarged. An important aspect we want to stress is that this way of operation achieves a greater use of surface water without need of artificial recharge.

Surprisingly enough, the possibility of regularly using more surface water in wetter periods has not been used very often. In many Mediterranean basins in Spain, besides the fields traditionally irrigated with prior rights, additional areas were irrigated with surface water in humid years. After the rapid increase of aquifer exploitation in the 1960s, they were integrated smoothly into the existing systems. So more surface water was used during wet periods, and more ground water was pumped during drought periods. In all cases the schemes were proposed and handled by the users. In other cases canals have been built by the hydraulic administration to substitute ground water for surface water in areas partly irrigated with ground water. In some cases surface water diversion is insufficient and varies from dry to wet years, so ACU is installed. More recently some of those existing practices in Valencia have been legally approved and additional alternative use schemes have been proposed.

The California Water Plan proposed a large-scale alternative conjunctive for the Central Valley that is the first and largest planned scheme of this type.

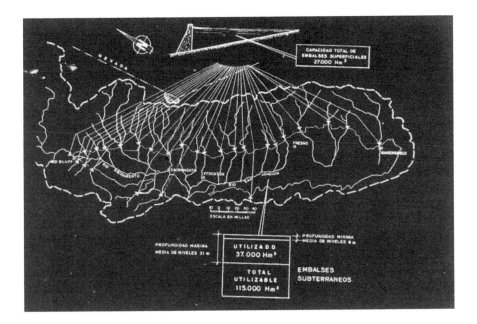

Figure 3.1 Alternate conjunctive use proposed in the Central Valley. California Water Plan 1957.

The total proposed storage between existing and proposed dams was 24,000 million cubic meters, and the used subsurface storage, considering the difference between forecasted highest and lowest ground water levels, was 37,000 million cubic meters (see Figure 3.1). Using that subsurface storage more surface water can be supplied without resorting to artificial recharge. Notwithstanding, supplementary use of artificial recharge was foreseen in the plan, but the proposal was not implemented as planned. Although we did not find a direct explanation for this change in the plan, it can be speculated that difficulties in the complex legal status from California occurred. Many individual projects, including dams and artificial recharge, have been built. Later, for many basins, "in-lieu recharge" has been applied to satisfy a demand of water when there exists a possibility of using surface water that cannot be stored (California State Department of Water Resources, 1957).

In the Mijares basin on the Mediterranean coast of Spain 60 km north of Valencia ACU is being practiced. There are three storage reservoirs: one upstream in the Mijares river with 100 million cubic meters of capacity, the second downstream in the main river, and the third in a nonpermanent tributary with 50 and 28 million cubic meters of storage respectively. The latter two reservoirs, built in karstic limestone, have important loses of water, on the order of 45 million cubic meters per year, which recharges the aquifer of La Plana de Castellón. The Mijares river also loses around 45 million cubic meters per year, which recharges the aquifer with a water table 20 to 40 meters below it. About one-third of the irrigated surface is supplied

alternatively with surface or ground water, depending on surface water availability in the river and stored in reservoirs. Traditional irrigated fields cover
one-third of the total irrigated area, which uses surface water, and the other
third and urban and industrial needs are covered exclusively with ground
water (Figure 3.2). When more surface water is available, aquifer recharge

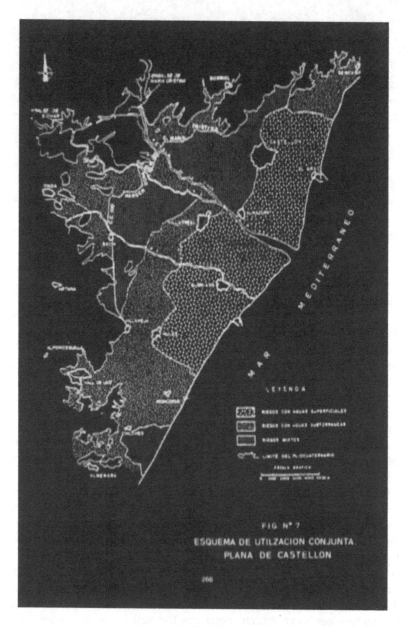

Figure 3.2 Alternate conjunctive use in La Plana de Castellón (Spain).

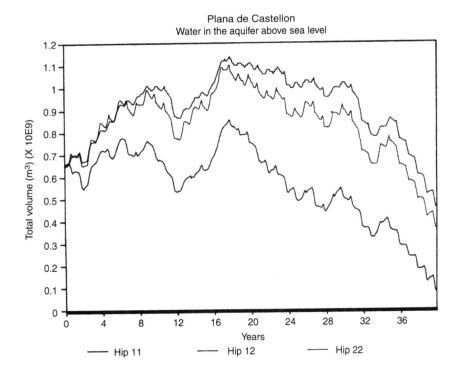

Figure 3.3 La Plana de Castellón aquifer. Change in storage for different use alternatives.

augments not only due to higher rainfall but also due to higher storage and river loses, as well as to recharge from some ephemeral streams flowing over the aquifer. The difference between high and lower values of water in storage in the aquifer can attain more than 600 million cubic meters, around four times the existing surface storage (Figure 3.3). That allows a large percentage of the average surface water in the basin to be used. Simulation showed that alternatives with a higher area irrigated alternatively with both surface and ground water could increase water availability slightly. Similar results are obtained for alternatives using artificial recharge as a big portion of the total water resources already captured.

A project to improve irrigation efficiency is currently under way in La Plana de Valencia. This improvement will largely diminish aquifer recharge and consequently its discharge to the Júcar river and to Albufera lake. That can produce negative influences over downstream surface water users and on the lake's ecology. Additionally La Plana de Valencia aquifer, although largely misused, became an important component in the regional water resources system. Being a component of an ACU scheme it is easily able to supply enough water in drought periods and implement other uses, including a local water transfer to the Alicante Province in the south. In the same system the Canal Júcar–Turia has been built to provide water to ground water irrigators. In fact

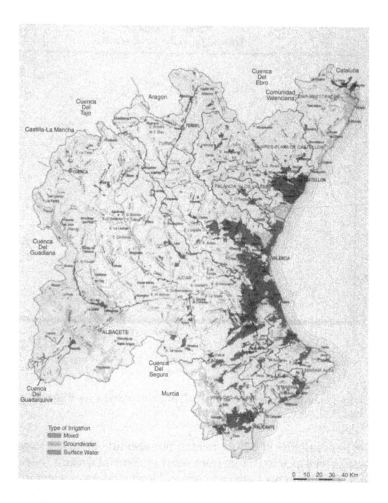

Figure 3.4 Alternate conjunctive use in the Júcar basin, Spain.

the higher altitude areas in the west continue to be irrigated with ground water. The eastern areas, on the right margin of the canal, use more surface water in wet years while pumping more ground water during dry ones. At the end of the 1991–1995 drought period the Júcar Basin Water Agency, joined with the regional Ministry of Agriculture, drilled and installed 65 large-capacity wells near the main canals in La Plana de Valencia area. They had barely started operating as the drought ended soon after wells were installed. The concept of ACU is used all around the region as can be seen in Figure 3.4 where irrigation areas utilizing surface water, ground water, and both sources are indicated.

ACU, as an alternative to building new dams, has been used to increase the capacity of the water supply systems of the Madrid metropolitan area. The existing capacity of wells has been increased up to 4 cubic meters per second, and additional increments have been foreseen. So assurance against drought is provided by increasing water supplies. Simulations of the conjunctive use show that a global increase of the annual firm yield is between two and three times each cubic meter of ground water pumped (Sánchez, 1986). The increase of yield comes mainly from a higher use of surface water in wet years.

Overexploited aquifers can be alleviated through conjunctive use with existing or projected surface water elements, although in some locations pumping patterns or their capacity have to be changed, as in the Campo de Dalias aquifer, jointly with the new Beninar dam in the Adra river in south-western Spain. Similar possibilities exist in many other schemes.

3.2.3 Stream-aquifer systems

The alternative use concept has been applied to alluvial and other small aqui-fers in conjunction with the rivers connected to them, constituting "stream-aquifer" cases. An important feature is that the mutual influences between river and aquifers are relatively more rapid that in other aquifers. Aquifer storage and flow cause a delay between well pumping and a decrease in river flow, because this river-aquifer interaction is of foremost importance. The specific delay depends on the distance from the pumped well to the river, the aquifer-river connection, and the aquifer geometry and diffusivity (ratio of transmissivity to aquifer storativity). Pumping during dry seasons increases water availability in an amount equal to the pumped quantities minus the effect of pumping on river flow. A part of the effect of pumping over river flows subsequently carries on over wet periods, when river flows are higher and demands are lower. Subsurface storage is created by ground water level descent as a result of aquifer pumping. After the pumping ceases, the depres-sion on ground water levels drops. Classic examples of aquifer-river conjunc-tive use system can be seen in the irrigation of the valleys of the rivers Arkansas and South Platte in the state of Colorado in the U.S. (Bredehoeft and Young, 1972; Morel-Seytoux et al., 1973). The South Platte is connected to an aquifer estimated to contain more than nine billion cubic meters of water, and the aquifer connected to the river Arkansas contains around 2.5 billion cubic meters of water (Heikkila et al., 2001). The same scheme is repeated in other aquifer-river systems of the central U.S. In the U.K. a very efficient use is being made of aquifer-river systems. Aquifers are constructed mainly in consoli-dated rocks, limestone, sandstone, and chalk. They are generally small in size, and their storage is less than in alluvial deposits, meaning that pumping has a relatively fast effect on river flow. Ground water is pumped and piped into certain rivers during dry periods to maintain adequate flows in them to meet supply and environmental requirements. This is also called "river augmenta-tion," and it is used systematically in a very efficient way in water planning in England and Wales (Skinner, 1983; Downing et al., 1974).

Artificial recharge is practiced even in areas where ACU is predominant. In the above mentioned South Platte aquifer some artificial recharge is under-taken to supplement stream flow. The fact that artificial recharge is done in the

South Platte and not in the Arkansas is due probably to the smaller width of the aquifer that would produce a faster return of recharged water to the river.

3.3 Comparison between artificial recharge and alternate use

When water is imported through large conveyance facilities, such as the water transfers in Southern California, Israel, and the Central Arizona Project, artificial recharge is the appropriate option, and alternative use makes no sense. In arid areas surface water is usually less important, and its variability is extremely high. Alternative conjunctive use loses some of its advantages, although it can be used favorably in some cases. Permanent rivers frequently exist in the wetter upper part of arid basins. Permanently flowing water and flood water can be used jointly with aquifers in dryer downstream reaches through artificial recharge or alternative use. In less arid environments, where alternative conjunctive use is being employed as in Spain, surface water flows exhibit a high temporal variability, but are not as sporadic as classic ephemeral streams in arid environments.

Each type of conjunctive use has its best application under different conditions of climate, geology, water supply availability, legal and regulatory environment, and economic development. The type based on artificial ground water recharge needs a more complex infrastructure for its successful operation. In less economically and technically developed countries the influence of artificial recharge operation and maintenance cost in the final cost of water could be too high for irrigation. Therefore, any doubts that the development of methods for enhancing natural aquifer recharge, or for lowering the cost of artificial recharge, should be investigated. Without any doubt the use of artificial recharge is compatible with ACU or other methods of enhancing the availability of water resources if the cost allows its use.

One additional point with artificial recharge is that it requires adequate technical operation and monitoring and permanent supervision. Furthermore, it cannot be implemented without well-identified users, the ability to pay for the operation and maintenance cost of recharge, and assurance that others will not pump the recharged water. This involves a high degree of institutional development that is far from being achieved in most countries. These difficulties hamper the development of large-scale artificial recharge projects in extensive irrigation districts unless they are operated and supported by governments.

Most cases of artificial recharge not used for water treatment or seasonal storage of potable water are found in the western U.S., particularly California. For our purposes here, directed toward developing countries, which impose a cost limitation, we excluded additionally artificial recharge sites used for water banking. In many of the remaining cases of artificial recharge, alternative use would have been economically competitive and probably much less expensive. The increase of sites where "in-lieu recharge" has been implemented is a clear indication of the interest of the alternative use concept.

Alternative use apparently faces possibilities and opportunities, particularly in wide-spectrum social and economic situations. In most Spanish basins it is possible to implement alternative use schemes. In any case, the possibilities depend on the variability of surface flows, aquifer storage, location and water volumes required by the different demands, aquifer situation and properties, and their relation to rivers. But as a general rule, advantages can be obtained whenever there exists an aquifer, a river with or without a dam or the possibility of building one, and unsatisfied demands for water. Users have promoted most of the alternative use schemes in the Mediterranean coast of Spain. They appear to work easily and without any major problems.

3.4   Other aspects and possibilities

3.4.1   Transformation of aquifer-river relationship due to ground water pumping

Heavily exploited aquifers can change their relation with a previously gaining river that is converted to loser. So the possibilities of storing water in the aquifer increases. A well-known case is in the lower Llobregat river, which became a loser after the aquifers where heavily pumped. La Plana de Castellón aquifer, previously mentioned, seems to have been draining into the Mijares river on the order of 20 million cubic meters at the beginning of the 20th century, and now the river recharges to the aquifer of the order of 40 million cubic meters. That situation is very common at many permanent rivers in the Mediterranean coast of Spain, where most rivers lose water, recharging the aquifers at the entry of the coastal plain; in many cases this reversal is produced by aquifer exploitation. This can be utilized to augment aquifer recharge, in some cases through adequate water releases and well-planned operation of surface storage.

3.4.2   Use of karstic springs

In Spain pumping in the aquifer to augment water availability for irrigation and urban water supply has regulated several karstic springs. In some cases wells have been located near the spring, in the proximity of existing canals or aqueducts used to transport the spring flow. In such cases the results of pumping wells is quick; pumping is implemented to augment the spring flow when natural flow is below water demand and after spring run-off dries out, but all water required must be pumped once pumping starts. Operating in this way implies that supply can be augmented well over the natural flow of the spring during the irrigating season as for urban or industrial needs. So the usually large variations of flow in many of those karstic springs can be accommodated to water demand. The use of the aquifer as a subsurface reservoir is very intuitive when the spring dries out. In many cases very high flows have been obtained in wells. Up to 1200 liters per second were obtained in two wells in the Los Santos river spring in Valencia. In the Deifontes spring near Granada in southern Spain, more than two cubic meters per second were provided for five 100-m deep wells. In other cases, the spring is a component of more complex schemes. So it happens in the Marina Baja water supply scheme having two dams, two aquifers, one being the El Algar spring, and treated water reused. Alternative use of ground water and surface water and the regulation of the El Algar spring by wells solved the acute supply problem suffered by a very important tourist area near Alicante in the Mediterranean coast of Spain. The two wells near the spring can each pump up to 400 liters per second and are used exclusively during dry periods. The underground storage provided by the aquifer during the large drought of 1990–1996 was estimated to be on the order of 40 million cubic meters, three times the existing surface storage. Another case where exceedingly high flows were obtained, 2.25 m³/s with five 100-m deep wells, is the regulation for irrigation purposes of the Deifontes karstic spring near Granada. There exist other additional possibilities in karstic areas in Spain of regulated springs that could be included in more complex conjunctive use schemes.

3.4.3 Alleviation of land drainage and salinization in irrigated areas and conjunctive use

In many irrigation projects aquifer recharge has increased due to water loses from conveyance and distribution systems in addition to infiltration surplus of applied water. Those increments in aquifer recharge can increase the potential for ground water development, and in arid zones has also produced drainage and salinity problems due to rising ground water levels. This is a customary problem of large surface irrigation projects in arid countries. The Planning Commission of the Government of India has recognized problems of water logging due to average water table rising, which is about 1 m per year on average in several schemes. So they suggested, in addition to enhanced water use efficiency, to increase ground water use jointly with canal water to augment supplies and prevent land deterioration. The total area affected by water logging due to both ground water rising and poorly controlled and inefficient irrigation was estimated in 1990 at 8.5 million h, with other estimates of 1.6 million h (Burke and Moench, 2000).

The drainage and salinity problems created in the Punjab plain in Pakistan have the same origin of surface water infiltration along the irrigation system of the Indus river and its tributaries. Irrigation started to be intensively developed in the late 19th century under British colonial rule. During the middle of the past century every year 25,000 h had to be abandoned and in 1960 2 million h, a total of 14 million h irrigated, were abandoned. The irrigated area is dominated by 43 big canals with a total length of 65,000 km, in addition to secondary and tertiary canals. The biggest 15 have flow capacities between 280 and 600 cubic meters per second. They are fed by several big dams — Mangla dam and Tarbela dam with 5.5 and 10.6 billion cubic meters of storage respectively. Most canals are unlined and have big losses that feed the huge aquifer below. Water levels rose 20 to 30 m, and up to 60 m in some places, in 80 to 1000 years. The problem has been intensively studied since the 1960s. The water resources group of Harvard University proposed to drill 32,000 high capacity wells to pump 70 billion cubic meters per year to lower the water table taking out to the sea the pumped salty water through lined canals and using the fresh ground water jointly with surface water to increase irrigation. A project to improve salinity and drainage conditions through groundwater pumpage, the Salinity Control and Reclamation Project (SCARP) was implemented. Since drainage projects do not have an immediate economic profit, most ground water pumped from wells was fresh water that was used to increase irrigation. In the same way, pumping of saline water and lining of canals to avoid the infiltration of salt water were not addressed. On the contrary, when salty water was pumped into a well it was blended with surface water to irrigate. So the salt balance of the aquifer increased instead of decreased. In some areas pumping and mixing of water of diverse salinities has increased the salinity erratically. Nevertheless, improvements of drainage and descent of soil salinity was quite important. Another important aspect not considered in earlier plans was the capacity of the private sector to get funding to develop ground water and drill depth high-capacity wells, with capacity triggered by SCARP realizations (Fiering, 1971; Burke and Moench, 2000; van Steenbergen and Oliemans, 2002). Some annalists argue that ground water overexploitation in Punjab exists, but information is not clear. In any case the target in heavily irrigated arid areas in the third world is to use existing aquifers, additionally recharged by return flow irrigation and by surface water infiltrated in the conveyance and distribution canals, jointly with surface water, while maintaining ground water levels below prescribed heads to contain salinity and drainage problems. Equally important is to control migration and perturbation of the more saline ground water bodies. So ground water quality can be maintained in addition to augmenting total water availability. Hydrogeological analysis and monitoring are needed in addition to the long-term simulation of ground water flow and salinity.

The same drainage and salinity problem exists in Egypt, northern China, and the Asiatic countries of the former USSR, where Kats (1975) suggested using jointly with surface water the estimated 25 billion cubic meter of water drained from irrigated lands. Losses at canals and distribution systems can be lowered with lining conductions, but if losses feed usable aquifers and conjunctive use is practiced, it can be more convenient to leave canals unlined, unless drainage problems exist and water losses contribute to maintain an excessively high ground water level (Task Committee on Water Conservation, 1981).

3.5 Conjunctive use potential in developing countries

Over the past 20 years many nations have increased ground water exploitation for agricultural irrigation purposes. Ground water resources have been underpinning the "green revolution" in many Asian nations. Access to ground water for irrigation purposes is making a very positive impact on subsistence and income for poor farmers, and in many cases it also reduces the need for the rural poor to migrate during droughts. Ground water use reduces agricultural risk and enables farmers to invest and to increase production. Some governments in developing countries have encouraged ground water development to meet the needs of rural populations as a mechanism for increasing their political popularity, regardless of the condition of aquifers (Foster, 2000; Shah & Ded Roy, 2002; Burke & Moench, 2000; Burke, 2002; Moench, 2003).

On the other hand, in many surface water irrigation projects, aquifer recharge has increased due to water losses from conveyance and distribution systems in addition to the infiltration surplus of applied water. Such increments in aquifer recharge can increase the potential for ground water development, and in arid zones they have also produced drainage and salinity problems owing to rising ground water levels. This is a frequent problem of large surface irrigation projects in arid countries. The Planning Commission of the government of India has recognized problems of water logging as a result of water table rising, which is about 1 m per year on the average in several schemes (Sondi et al., 1989). Consequently they have suggested, in addition to enhanced water use efficiency, increasing ground water use jointly with canal water to augment supplies and prevent land deterioration. The total area affected by water logging, due to both ground water rising and poorly controlled and inefficient irrigation, was estimated in 1990 to be 8.5 million h, while other estimates indicate 1.6 million h (Burke and Moench, 2000).

Conjunctive use can undoubtedly increase water availability in many existing or planned schemes where both surface water and ground water resources exist. In some cases conjunctive use is claimed to be applied but only when advantage is taken of the conveyance, distribution, or storage capacity of its components and the system is properly operated can it be considered as conjunctive use. But as a general rule, advantages can be obtained whenever there exists an aquifer, a river with or without a dam, or the possibility of building it. The advantages could be increasing water availability or alleviation of aquifer overexploitation.

3.6 *Analysis of conjunctive use systems*

It is important for the design and operation of conjunctive surface and ground water resources systems to adequately evaluate Alternative Conjunctive use ACU's performance. Good performance is also required to convince all stakeholders of ACU's effectiveness in water-related problems: governments, water agencies, and other public administrations and users.

As discussed above, the design and management of conjunctive surface and ground water resources systems have a higher degree of complexity than systems of surface water or ground water alone. In any case, a good system analysis practice is recommended to obtain good results, and in conjunctive use this is a necessity. Thus, some particular aspects of the analysis have to be carefully considered:

- The assessment of surface and groundwater resources has to be done jointly. It is very common to make a separate assessment of the resources by building and calibrating more or less sophisticated models for surface hydrology and more or less sophisticated models for ground water hydrology. Both types of models are usually calibrated in order to best reproduce the observed values at some locations of surface flow in the first case, and of piezometric levels in the second case. Due to the great number of parameters used in the models, this separate calibration does not guarantee that the interactions between the surface system and the aquifer are well captured. In fact, according to the Spanish experience, coupling of such separately calibrated models often produces incoherence in the resulting joint models. The situation is much worse when the natural surface and ground hydrological regimes are disturbed by man's activities. In such cases, the duplication of natural flows needed for the calibration of surface hydrology models requires the duplication of ground water interactions, which require a ground water model, which in turn might need the values of the surface flows as inputs. Consequently, joint modeling and calibration is needed, with models oriented to capture in the best possible way the interactions between both subsystems. For conjunctive use modeling, this is more important than achieving a better tuning to other responses.
- The analysis of conjunctive use alternatives has to include streams, reservoirs, canals, aquifers, and flow interchanges between ground and surface water, in addition to water supply facilities for different uses. Consequently, it has to be conducted on a regional scale, with the basin scale being the most adequate in many cases.
- Conjunctive use is a matter of management. Therefore, operating rules are important components of the alternatives. The same set of structural facilities can produce very different yields depending on the operation of the system, so they have to be explicitly incorporated in the analysis and be realistic enough to be applied in real life. There are many political, historical, sociological, and cultural factors that can impede the application to the real world of otherwise perfect operating rules.
- Modeling of ground water components must be as detailed as needed for the purpose, yet emphasizing surface-ground water interactions. The use of the tools must be facilitated through the possibilities provided by the modern concept of computerized decision support systems. State-of-the-art models and methodologies should be put in the hands of the real-world practitioners and decision makers in order to study such complex systems for a large number of alternatives.

If these conditions are met, the more convenient conjunctive use strategies can be devised, including components, the design of the infrastructure needed, and the operating general guidelines. For the latter steps, it is advisable to achieve the effective application of the conjunctive use strategies and continuous monitoring of the water bodies. This is better accomplished through the users because the users associations can prevent individual objectives from becoming community interests.

It is also necessary to have tools that help in the decisions in regular basic management of the conjunctive use, in order to adapt the general operating guidelines to the existing hydrological circumstances. From the information provided by the continuous monitoring of the water resources system and the information on the water requirements, future scenarios for the short to medium term (e.g., some months) can be analyzed, and the risks of affording shortages can be evaluated. Then, anticipated measures can be adopted to mitigate the effects of such an operational drought.

3.7   Methods of analysis

As a semiarid country, Spain is concerned with the use of surface water and ground water resources and has acquired experience in the analysis and management of conjunctive use systems by applying advanced decision support systems (DSS). AQUATOOL is a generalized DSS developed during the past 20 years at the Universidad Politécnica de Valencia (UPV), to optimize and simulate complex systems including conjunctive use. This method, which has been applied in many Spanish basins, can handle several dams, aquifers, and demand areas including rivers, canals, aqueducts, and aquifer-river interactions, and it can tackle the most common nonlinear situations. It has been designed to help decision makers analyze complex systems in order to answer specific questions, facilitating the use of a set of models and databases in an interactive way in a user-friendly control framework. One of its capabilities is the possibility of using a methodology that solves the space discretized ground water flow equation allowing a very efficient integration of an aquifer model in the simulation of complex systems with conjunctive use.

The eigenvalue method for aquifer simulation solves the same flow equation that classical methods do. Space discretization is indicated as infinite differences or finite elements methods; conversely, time is continuous, and solutions are given continuously on time. Therefore, their accuracy is not less than that provided with classical discrete time solutions. Its most interesting advantage is that it explicitly and easily transforms the current state of the system in a state vector from which piezometric heads, vector flows, or surface water ground water interchange flows and can also be very easily obtained. And that needs to be made only for a few points and times of interest. Computer work needed to obtain the basic eigenvalue solution of the problem makes this approach not competitive with normal modeling needs, but it is especially adequate for analyzing some alternatives with important accumulated simulated periods of time, as usually occurs in ground water management problems. As the influence function method, it applies to linear systems with no temporal changes in transmissivity or storage coefficients.

3.8 Conclusion and recommendations

Due to the usually very high investment associated with dams and canal building and the present trend toward ground water development, there is a great potential for conjunctive use in many developing countries in arid and semiarid regions. Increase of ground water pumping during droughts has been a common practice all over the world for decades; and it is expected to continue. In many cases ground water pumping complements surface water availability but usually to a limited extent. It is a limited stage of conjunctive use. In such cases surface water surplus can often occur during wet years, while ground water resources are extensively exploited. The logical extension is not only to mitigate droughts by augmenting ground water pumping, but also to try to use as much surplus surface water in wet years as possible and proceed to an integrated ACU. The rationale behind this strategy is that advantages can be obtained for the aquifer whose stress diminishes and through a higher use of surface water resources during wet years. Both gains are obtained without augmenting surface storage and without the need of artificial recharge.

In arid countries, due to surface water irrigation, return flow increases aquifer's recharge, thus increasing ground water levels. Drainage and salinity problems often arise (e.g., India, Pakistan, China, Egypt, Asian countries of the former USSR, Argentina), and sometimes millions of hectares are affected and abandoned. Ground water pumping can solve or attenuate drainage and salinity problems, but it is only practiced in a few cases and in a limited way. In some cases, more dams have been built, thereby exacerbating drainage problems. Conjunctive use could be used both to increase water availability and to treat drainage problem. The case of the Indus irrigation scheme in Pakistan is one of the most enlightening and interesting ones.

As a general rule, conjunctive use can help whenever an aquifer and a river (with or without a dam) coexist. Improvement of many schemes can be achieved rather inexpensively and quickly through ACU, but adequate institutional and social changes would be needed in most cases. It can be concluded that a conjunctive use is an essential aspect of integrated water resources. Among the conjunctive use schemes, ACU is very attractive for semiarid regions of developing countries. On the other hand, the analysis for implementing conjunctive use has to be carefully performed; management is perhaps the most crucial single factor, and a strong political will is needed to implement such systems. It was obvious from the experiences in various countries that some level of organization must be provided for an effective application of conjunctive use. One important point to stress is that every improvement should be made according to the users needs and cultural behavior. We are confident that enhancement of the role of water users can work in most cases and the instauration of Water User Associations will work to make irrigation systems more equitable (Shah et al., 2000; Rao 2000).

Some of the beneficial aspects we can expect from the conjunctive use of surface water and ground water are:

- Alleviation of drainage and salinity problems
- Alleviation of aquifer overexploitation
- Alleviation of sea-water intrusion
- Higher reliability
- Smaller infrastructures
- Increase in economic optimization

However, it must be said that in globally overstressed basins (e.g., Mexico City, Segura basin in Spain) few quantitative gains are possible; only reallocation of resources is feasible.

References

Bredehoeft, J. D., and Young, R. A. (1983). Conjunctive use of groundwater and surface water for irrigated agriculture: risk aversion. *Water Resour. Res.* 19(5), 1111–1121.

Burke, J. J. (2000). Groundwater for irrigation: Productivity gains and the management of hydro-environmental risk. In M. R. Llamas and E. Custodio (Eds.), Intensive use of groundwater: Challenges and opportunities. Netherlands: A. A. Balkema.

Burke, J. J., and Moench, M. H. (2000). *Groundwater and society. Resources, tensions and opportunities.* New York: United Nations Department and Social Affairs and Institute for Social and Environmental Transition.

California State Department of Water Resources. (1957). The California water plan. Bull. 3.

Custodio, E., Isamat, F. J., and Miralles, J. M. (1969). Twenty-five years of groundwater recharge in Barcelona (Spain). In *International symposium on artificial groundwater recharge.* Paper 10. DVWK. Bull. 11 (pp. 171–192). Hamburg-Berlin: Verlag Paul Parey.

Downing, R. A., Oakes, D. B, Wilkinson, W. B., and Wright, C. E. (1974). Regional development of groundwater resources in combination with surface water. *J. Hydrology* 22, 174–177.

Fiering, M. B. (1971). Simulation models for conjunctive use of surface and ground water. Seminar of Ground Water, Granada, Espagne, FAO-Spanish Government: 1–25.

Foster, S. (2000). *Sustainable groundwater exploitation for Agriculture; current issues and recent initiatives in the developing world.* Papeles del Proyecto Aguas Subterráneas. Serie A. Uso intensivo de las aguas subterráneas. Fundación Marcelino Botin.

Heikkila, T., Blomquist, W., and Schlager, E. (2001). Institutions and conjunctive water management among three western states. Tenth Biennial symposium. *Artificial Recharge of Groundwater.* Tucson, A2.

Kats, D. M. (1975). *Combined use of groundwater and surface water for irrigation. Soviet hydrology. Selected papers.* American Geophysical Union, International Water Resources Association and American Water Resources Association: 190–194.

Llamas, M. R., and Custodio, E. (Eds.). (2003). *Intensive use of groundwater: Challenges and opportunities.* Netherlands: A.A. Balkema.

Moench, M. H. (2003). Groundwater and poverty: Exploring the connections. In M. R. Llamas and E. Custodio (Eds.), *Intensive use of groundwater: challenges and opportunities.* Netherlands: A. A. Balkema.

Morel-Seytoux, H. J., Young, R .A., and Radosevich, G. E. (1973). *Systematic design of legal regulation for optimal surface groundwater usage-PHASE 1.* Colorado State University, Environmental Resources Centre, Report 53.

National Research Council (NRC). (1997). *Valuing groundwater. Economic concepts and approaches.* Washington DC: National Academic Press.

Pyne, R. D. G. (1989). Aquifer storage recovery: A new water supply and groundwater recharge alternative. *Proceedings of the International Symposium on Artificial Recharge of Ground water.* A. I. Johnson and D. J. Finlayson (Eds). American society of civil engineers. New York, 107–121.

Rao, P. K. (2000). A tale of two developments of irrigation: India and USA. *Int. J. Water* 1(1), 41–60.

Rivera, A., Sahuquillo, A., Andreu, J., and Mukherji, A. (2004). Opportunities of conjunctive use of groundwater and surface water. In *groundwater extensive use.* A. Sahuquillo, J. Capilla, L. Martinez–Cortina, and X. Sanchez-Vila (Eds.). International association of Hydrologists. Selected papers. Rotterdam. A. A. Balkema Publishers.

Sahuquillo, A. (1983). An eigenvalue numerical technique for solving unsteady groundwater models continuously in time. *Water Resour. Res.* 12, 87–93.

Sahuquillo, A., and Lluria, M. (2003). Conjunctive use as potential solution for stressed aquifers: Social constraints. In M. R. Llamas and E. Custodio (Eds.), *Intensive use of groundwater: Challenges and opportunities.* Netherlands: A. A. Balkema.

Sánchez, A. (1986). *Criterios para la evaluación del coste del agua en Madrid. Jornadas sobre la Explotación de Aguas Subterráneas en la Comunidad de Madrid.* Canal de Isabel II.

Shah, T., and Ded Roy, A. (2002). Groundwater socio-ecology of India. In M. R. Llamas and E. Custodio (Eds.), *Intensive use of groundwater: Challenges and opportunities.* Netherlands: A. A. Balkema.

Skinner, A. (1983). Groundwater development as an integral part of river basin resource systems. In *Groundwater in water resources planning.* IAHS Publication 142.

Sondi, S. K., Rao, N. H., and Sarma, P. B. S. (1989). Assessment of groundwater potential for conjunctive water use in a large irrigation project in India. *J. Hydrology* 107, 283–296.

Task Committee on Water Conservation of the Water Resources Planning Committee of the Water Resources Planning and Management Division. (1981, March). Perspectives on water conservation. *J. Water Resour. Plann. Manage. Div.* ASCE 107(1).

UK Groundwater Forum. (1998). *Groundwater our hidden asset.* Wallinford, UK: R. A. Downing.

Van Steenbergen, F., and Oliemans, W. (2002). *A review of policies in groundwater management in Pakistan 1950–2000. Water Policy.* The Netherlands: Elsevier.

chapter four

Optimization model for the conjunctive use of conventional and marginal waters

Francesca Salis, Giovanni M. Sechi, and Paola Zuddas
University of Cagliari, Italy

Contents

4.1   Introduction

There have been significant computational advances in the development of optimization techniques that are able to consider the inclusion of special risk-related system performance criteria within the analysis of different hydrologic and demand scenarios. Advances have also been made when focusing on the water system domain and the complexity of modeling tools.

Nevertheless, mathematical optimization procedures available for large systems are still not able to deal with all of the complexities and nonlinearities of the real world that can be easily incorporated in a simulation model. Optimization procedures can also be constructed to solve efficiently the real problem and are an adequate approximation to it, and simulation can greatly narrow the search (Loucks et al., 1981). Moreover, optimization results obtained by solving an adequately adherent model can be seen as the management reference-targets for simulation since they can be considered as obtained by an ideal-manager having perfect knowledge of sources and demand behavior in the assumed time horizon. Reflecting upon tools for water management, in one view to the future, Simonovic (2000) identified two paradigms that will shape tools for the future modeling of water management. The complexity paradigm states that water problems in the future are going to be more complex and will need to take into consideration more extended domains such as environmental and social impact, population growth and needs, water quality indicators, a longer temporal horizon, large-scale water problems, etc. The uncertain paradigm, on the other hand, is the increase in all elements of uncertainty in time and space. Uncertainty in water management has been divided as reported into two basic forms: uncertainty caused by hydrologic variability and uncertainty due to fundamental lack of knowledge.

The main aspect of improving the practical utility of optimization in water resources revolves around better packaging of associated computer software. Probably, at the moment, this is the primary requirement needed to convince decision makers to accept optimization as a real problem-solving tool. As recently pointed out in Nicklow (2000), decision makers are more likely to accept optimization if they are comfortable with their abilities to employ the computational model and if the model has an interactive graphical user interface (GUI). Consequently, adequate consideration should focus on providing user-friendly optimization software for water systems applications, and additional emphasis should be given to this requirement.

Following the above-mentioned aims, in this chapter we describe a general-purpose optimization tool (Sechi and Zuddas, 2000, 2002) characterized by considering a user-friendly interface. The proposed package, named WARGI (Water Resources System Optimization Aided by Graphical Interface), was developed to consider water quality indices under the name Water Resources Management Under Drought Conditions (WAMME, 2002) European project in order to improve opportunities in real cases for using optimization, aided by a graphical user interface in water resource systems modeling. As will be illustrated in this chapter, the package has been developed in a scenario-modeling framework and considers the possibility of conventional and marginal water utilization for solving large-scale water system optimization problems.

As is well known, a complex regional water supply system consists of several components such as reservoirs, conveyance works, treatment plants, pumping stations, hydroelectric power plants, etc. The system may include several competitive water uses (urban, irrigation, hydroelectric, recreational, etc.), and the optimization model must take into account management alternatives as well as design problems, with the aim of reaching an optimal utilization of the resource.

Optimal configurations of the water resources system will guarantee an adequate level of reliability of water supply for different uses and provide management assessment criteria to be adopted by the water authorities. At the same time emergency plans should be prepared to reduce the consequences of drought events by allowing the interconnection of water systems and so on. For this purpose we need to examine very different and sufficiently detailed scenarios as efficiently and rapidly as possible.

In the WAMME project, the overall goal is to increase the scientific background and to develop technological tools for improving water resource management and environmental control in drought-prone Mediterranean regions. In order to develop strategies for identifying the role of different project and management alternatives in mitigating drought impacts in complex water systems, including the conjunctive use of conventional and marginal waters, WAMME identified an organized framework considering the following phases:

1. To develop an optimization model for the conjunctive use of conventional and marginal waters incorporating synthetic water quality indices;
2. To develop a simulation-based decision support system (DSS) for the management of integrated water resources systems focused on drought prevention and mitigation, which could help the decision makers to face drought risk in the Mediterranean regions;
3. To verify the usefulness of the developed optimization model and DSS tools using multicriteria techniques to identify the preferable mix of long- and short-term measures for drought impacts mitigation.

In this multiphase technique we need to resort to mathematical programming, applied to an optimization model, followed by a simulation testing process that limits recourse to heavy computational procedures by reducing the gap between the solutions of the first two phases. In the optimization phase we can examine a mathematical model representing the real physical problem simplified and consequently reducing its level of adherence to reality. Therefore, in the simulation phase we can only examine reduced scenarios from the dimensional and temporal points of view. On the other hand, if the optimization model remains adherent to the real problem, in a deterministic framework mathematical programming results give us the best management (in terms of water flows in the system) obtained by an ideal manager having previous knowledge of input and demands behavior in the system. These results can give us a measure of the "goodness" of the simulation-based DSS, at least for a reduced deterministic test cases set.

It is therefore crucial to be able to solve optimization models efficiently and with an adequate level of adherence in order to reduce the simulation phase computational effort and to have an objective measurement of DSS efficiency.

When facing droughts, the problem of the optimal dimension and management of the water resource system and the related optimal flows configuration should take into account additional constraints and costs in the objective function (OF) given by the criteria that it should operate satisfactorily during periods of drought. Particularly the vulnerability of the system should be considered when dealing with water resource shortage risk in the water system management. The vulnerability expresses the severity of drought in terms of its consequences. The consequences of drought are generally expressed by a loss (cost) function, and the measurement estimating the severity of drought is given considering cost functions, weighing more the shortage flows as the severity of the drought event increases.

In the optimization model, drought vulnerability will be examined considering a generalized OF expression using a standardized shortage to define the expected losses. In this way the problem can be expressed using a linear programming (LP) approach. Other approaches lead to a quadratic programming (QP) model.

Reliability (probability that the system is of a satisfactory state) and resiliency (recovery time from failure) indices can be evaluated in aposterior analysis. The ever increasing importance of problems related to water has created the need to improve knowledge concerning the phenomena involving water quality, in particular, when water resources are derived from surface water and artificial reservoirs as is the case in the main Italian islands (Sicily and Sardinia) supply systems. In a simplified frame, for river streams, reservoirs, and ground water resources, the problem of water quality can be studied considering the general environmental characterization of the water bodies, just as the recent Italian legislation has done.

On the basis of available measurements of basic macrodescriptors, surface water, ground water, and lakes can be divided into five classes; the first one should be attributed with the best environmental characterization of the water bodies, the last with the worst one.

The ecological state attribution derives from chemical and biological parameters. Measurement methodologies and attribution rules have been extensively described in the recent Italian legislation.

As will be described in the following paragraphs, using this simplified method, in the optimization model WARGI we will consider the potential use of water sources (conventional and marginal) and the technological costs associated with water treatment as a function of the ecological state attribution of the water bodies.

4.2   Identification of the optimization algorithm

Models describing the planning and management optimization problem of water resource systems show a special structure that suggests some specialized approaches in order to overcome the serious computational problems due to the large size of the models for complex systems.

Taking into consideration different physical situations, some specific optimization models together with algorithms exploiting the mathematical structure of the problem will be examined in the following paragraphs.

4.2.1   Modeling approaches and software tools

The physical water system can be represented by a direct basic graph in which nodes represent sources, demands, reservoirs, etc., and arcs represent the activity connections between them.

A "basic graph" can be deduced from the schematic representation of the physical water system referred to a single-period. This is a static representation of the system without taking time evolution into account.

In some cases the correspondence between the physical and graphic components is not so close; at times it is even nonexistent. In fact, it may be convenient to add "dummy" nodes and arcs to represent not only physical components but also events that may occur in the system. For example, dummy arcs can represent a shortage caused by meeting the request of the demand nodes and can prevent unfeasible solutions. This representation points out possible deficits in the system and, as a consequence, the need to change the dimensions of the works or, alternatively, to use external water resources. With each node we can associate a supply or a demand representing an input flow, such as a hydrological inflow to a reservoir, or an output flow, such as a water request from a demand center.

Similarly, with each arc we can associate a functional or technological activity such as for pipelines, power stations, hydroelectric power plants, etc., and the transfer costs and bounds for each flow. In water resources system analysis we also need to examine the evolution of flow values in time. This is done by extending the analysis to a time horizon sufficiently wide to describe the functionality of system components and the cost-benefit performance, which can also reach a significant representation of the variability of hydrological inflows and water demands in the network nodes.

In our analysis we divide the total time horizon in time steps (periods) usually taken as equal to one month. In our Mediterranean climate, which has a large annual rainfall variability, the total time horizon frequently needs to be extended to several decades. In the Sardinian Island Water Plan carried out by the Regional Authority the time horizon has been taken as equal to 54 years.

In a multiperiod analysis of the water supply system we must consider surface reservoirs as well as ground water resources and other storage activities that can transfer water to subsequent time periods to avoid shortage.

As a support to multiperiod planning analysis we can construct a "multiperiod graph" **R** using the "basic graph" as a multiplicative module reproduced as many times as are the periods considered, linking the copies by interperiod arcs between the storage nodes. Flows in these arcs represent the volume stored at the end of each period available for use during the next period. The support network allows the use of highly efficient data structures, reaching a significant reduction in computer storage and computational time during data input and processing. Modularity, moreover, allows for the automatic construction of the multiperiod network and of data generators.

Referring to a "static" or single-period situation, we can represent the physical system by a direct network (the "basic graph"), derived from the physical sketch. Figure 4.1a shows a physical sketch of a simple water system. In the figure nodes maintain the shape of the common hydraulic notation in order to recall the different functions of the system components. Nodes could represent sources, demands, reservoirs, ground water, hydroelectric power station sites, etc. Nodes corresponding to storage possibilities represent the memory of the system as they can store the water resource in a period to transfer it in a successive period. A dynamic multiperiod network (the "multiperiod graph") can be derived by replicating the basic graph for each period ($t = 1, \ldots, T$) to support the dynamic problem.

We connect the corresponding reservoir or ground water nodes for different consecutive periods by additional arcs (the "interperiods arcs") carrying

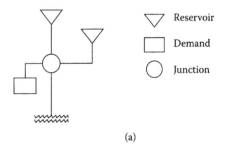

(a)

Figure 4.1a Sketch of water system.

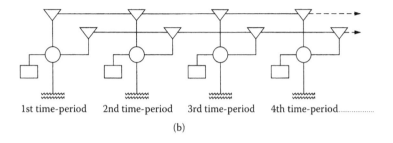

1st time-period   2nd time-period   3rd time-period   4th time-period..................

(b)

Figure 4.1b Segment of the dynamic network.

water stored at the end of each period. Figure 4.1b shows a segment of a dynamic network generated by the simple basic graph of Figure 4.1a. Reservoir nodes (symbol ∇ in the figure) are connected by interperiods arcs so they can store and transfer the resource in time. Demand nodes (symbol in the figure) correspond to a general requirement of using and consuming the resource. Confluence nodes (symbol O in the figure) allow the resource to pass without consumption.

The correspondence between physical and network components are not that close; it is even a sham at times. Figure 4.2 shows the dynamic multi-period graph, corresponding to that of Figure 4.1b, including dummy nodes and arcs marked with a dot. The basic graph is in the frame. The dummy node **U** represents a possible "external system" acting as a supposed source or sink. In this way each arc (**i,U**) represents spillway from reservoir nodes **i**; each arc (**U,i**) represents a supposed additional flow in case of shortage in order to meet the request in the demand nodes **i** and prevent solutions that are not feasible.

Flow on arcs (**U,i**) point out possible deficits in the system and the necessity of modifying the dimensions of the works or, alternatively, of making recourse to real external water resources.

In planning studies involving water systems characterized by a high seasonal and annual variation of hydrological inflows, such as in Mediterranean

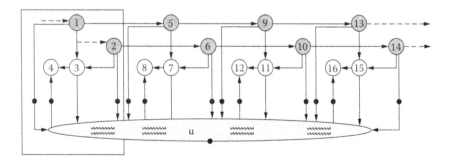

Figure 4.2 Dynamic multiperiod graph with dummy nodes and arcs.

countries, hundreds of replicas of the basic graph need to be considered, and this leads to a very large network model. For a given configuration of the system, and therefore for fixed requirements, resources, and work dimensions, the problem can only be viewed in an operational context, and a pure network flow model can adequately represent the performance of the system. In this way a network flow algorithm retrieves the best flow configuration, as if an ideal manager of the water system should take decisions knowing the time sequence of inflows and demands beforehand.

The high efficiency of network flow programming has been well known since the 1960s (Ford, Fulkerson, 1966) and has been tested extensively since the end of the 1980s (Ahuja et al., 1993), (Kennington, Helgason, 1980). Concerning water-resource systems, a comparison between commercial and public domain network flow codes (RELAX, NETFLOW, EASYNET, etc.) was performed (Sechi and Zuddas, 1998) to evaluate application possibility and performances.

The formulation of a model is characterized by the usual operative constraints and by the determination of the optimal scaling of supplementary works that allows for the reduction of the system shortage to predefined acceptable thresholds and modifies the pure-network shape of the model. Constraints that describe the links between project variables and operative variables are in this case also present, as well as constraints that represent the control on the deficit arcs ("shortage flows"). These constraints can determine, for example, that flows on deficit arcs do not exceed prefixed values, or they can impose limits on the sums of them.

Nevertheless, in the model size a significant part of the constraints is represented by the flow continuity equations and by predefined lower and upper bounds on the flows. In order to exploit the model structure and the performances of pure network programming, an expansion technique has been proposed (Sechi and Zuddas, 1995) that interacts between the primal and dual mathematical optimization model. This kind of approach is very useful in formulating a trade-off between the dimension of water works, the reliability of the system, and the prediction of severity in demand short falls.

Regarding a more general linear programming model, the most efficient state-of-the-art software tools (like CPLEX and X-PRESS) have been compared, when applied to water resources system optimization. Moreover, the hypergraph approach has been considered and compared (Sechi and Zuddas, 2001) by solving the design and management problem with the ordinary LP model, using the commercial CPLEX package (CPLEX, 1993) and with the HySimpleX code. The results obtained are very promising and show that HySimpleX can be adopted competitively for water resource design problems.

4.2.2 Optimization under uncertainty: The scenario optimization

Water resources management problems with a multiperiod feature are used in association with mathematical optimization models that handle thousands of constraints and variables depending on the level of adherence required in order to reach a significant representation of the system.

These models are typically characterized by a level of uncertainty concerning the value of hydrological exogenous inflows and demand patterns. However, inadequate values assigned to them could invalidate the results of the study. When the statistical information on data estimation is not enough to support a stochastic model or when probabilistic rules are not available, an alternative approach could be applied, that of setting up the scenario analysis technique. This is a general-purpose modeling framework to solve water system optimization problems under input data uncertainty. Scenario optimization is an alternative to the traditional stochastic approach, which is used to reach a "robust" decision policy that should minimize the risk of wrong decisions. This approach leads to a huge model that includes a network submodel for each scenario plus linking constraints, which must be treated with specialized resolution techniques.

In the proposed approach, the problem is to be expanded on a set of scenario subproblems, each of which corresponds to a possible configuration of the data series. Each scenario can be weighted to represent the "importance" assigned to the running configuration. Sometimes the weights can be viewed as the probability of occurrence of the examined scenario. A "robust-barycentrical" optimization solution can be obtained by a procedure that minimizes the distance between subproblems optima.

The model is usually defined in a dynamic planning horizon in which management decisions have to be made either sequentially, by adopting a predefined scenario independently, or by following different scenarios in a "scenario-tree" context since the data characteristics change as described in the next paragraph. The scenarios aggregation into a tree must be performed following the basic "implementable principle" or "principle of progressive hedging": "If two different scenarios are identical up to stage r on the basis of the information available about them at stage r, then the variables must be identical up to stage r" (Rockafellar and Wets, 1991). This condition guarantees that the solution obtained by the model in a period is independent of the information that is not yet available; in other words model evolution is only based on the information available at the moment, a time when the future configuration may diversify.

When the set of synthetic hydrological sequences has been generated, the principle of progressive hedging is performed by bundling the sequences to build the scenario tree.

Data defined for the deterministic model are required for each scenario in the chance model plus the further data:

G set of synthetic hydrological sequences (parallel scenarios)

w_g weights assigned to a scenario $g \in G$

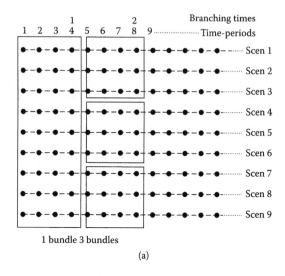

Figure 4.3a Set G of nine parallel scenarios.

Figure 4.3a shows a set of nine parallel scenarios before aggregation. Each dot represents the system in a time-period. Figure 4.3b shows an example of the scenario tree derived from the parallel sequences.

To perform scenario aggregation a number of stages are defined, where stage 0 corresponds to the initial hydrological characterization of the system up to the first branch time period. In the scenario tree this represents the root. In stage 1 a number, b_1 (3 in the figure), of different possible hydrological

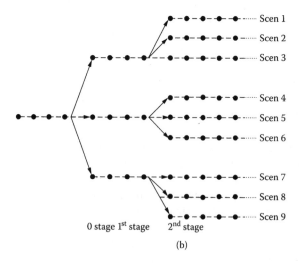

Figure 4.3b Scenario-tree aggregation.

configurations can occur; in stage 2 a number, b_1 *b_2 (9 in the figure), can occur, and so on and so forth.

The figure represents a tree with two branches: the first branching time is the 4th time period; the second is the 8th period. In time periods that precede the first branch, all scenarios are gathered in a single bundle, and three bundles are operated at the second branch. The zero bundle, includes a group of all scenarios; in the 1st stage 3 bundles are generated including 3 scenarios in each group, while in the 2nd stage the 9 scenarios run until they reach the end of the time horizon.

Finally, the main rules adopted to organize the set of scenarios are:

> *Branching*: to identify branching times **t** as time periods in which to apply bundles on parallel sequences, while identifying the stages in which to divide the scenario horizon.
> *Bundling*: to identify the number, $\mathbf{b_t}$, of bundles at each branching-time.
> *Grouping*: to identify groups, $\mathbf{G_t}$ of scenarios to include in each bundle.

In this way the graph grows in size, on increasing the possible branches, and each root-to-leaf path represents a particular scenario.

Once scenarios have been generated, some general checks must be performed to test their statistical properties: among others, a stationary test on mean and variance in order to check process changes over time; an independence test, to look for possible relations or for a trend among subsequent stages; a time and space correlation test; etc.

4.2.3 *Hydrologic series generation for scenario optimization*

Scenario optimization can be used to treat uncertainty concerning the value of hydrological exogenous inflows and demand patterns. Nevertheless, in dealing with optimization under drought conditions, this approach has been used more frequently concerning the variability in hydrological inputs. To avoid an excess of complexity when using the WARGI optimization tool, it was decided to adopt separate preprocessors to prepare hydrological scenarios to be used in the scenario optimization phase.

The data preprocessor that builds scenario sequences follows procedures that can be developed using different approaches. Therefore, the scenarios are to be viewed as a set of synthetic hydrological series obtained from historical samples applying time-series modeling procedures. Mainly, three approaches could be performed to generate the series:

- The first refers to autoregressive (AR) and autoregressive with moving average terms (ARMA) models to generate synthetic hydrological series
- The second to a Monte Carlo (MC) generation scheme
- The third to Neural Network (NN) techniques

Synthetically, the AR and ARMA models should be able to reproduce the most important statistical characteristics of observed time series generation.

Generally this technique first requires a normalization and standardization phase. In WARGI applications, a spatial desegregating routine divides the hydrological series into two families: principal (or independent) and secondary (or dependent). Using the normalized and standardized principal series, a hydrological series has been generated using an AR(p) or ARMA(p,q) model that can be selected for generation; for a secondary series a transfer model has been used for the generation of synthetic series.

An example of the MC approach used for synthetic runoff series generation is given in a WARGI application illustrated in a following paragraph. The MC procedure has been used for the generation of synthetic runoff series, which rescaled the historical series in order to impose a new mean and variance with respect to the historical values. Generally this generation procedure requires a preliminary definition of time-period clusters on hydrological data in order to avoid autocorrelation components, and it synthetically consists of the following steps: (1) the random generation of meteorological characterization at each cluster; (2) the generation of hydrological data from predefined sets of clusters; (3) the addition of noise components to improve the statistical fitness.

The NN approach for scenario generations, as required by WARGI, has been developed in recent papers (Cannas et al., 2000, 2001, 2002) using both the classical multilayer perceptron scheme and the locally recurrent NN scheme. In any case, a first sensitivity analysis was carried out to evaluate the best fitting NN configuration, the number of nodes in the layers, and the number of iterations to be used in model training. Subsequently (in the testing phase) the hydrological series was generated.

4.2.4 Considering water quality conditions in water system optimization

The ever increasing important problems related to water scarcity have resulted in the need to improve the knowledge about the phenomena involving water quality; in particular, when water resources are derived from reservoirs, as is to be found in the Sardinia (Italy) supply systems.

The indications given by Italian legislation (Law 152/1999) for classification are different for each water body. The common characteristics concern the discovery of two groups of parameters: one that is compulsory, the other that concerns dangerous substances. The choice of dangerous substances to be examined is made by the local authorities on the base of anthropic pressure factors existing in the hydrographic basin, based on limit values in the norm 76/464/CEE, which allows for the evaluation of the chemical state.

Among the compulsory parameters, the law points out a limited number of parameters called "macrodescriptors" used for the classification of the ecological state of water bodies. The other parameters serve to give support information on the principal characteristics of the water bodies or on the entity of the loads transported.

The macrodescriptors should allow for the precise and stable measurement of the load caused by anthropic presence and diffused activity: they therefore contain all those parameters that measure nutrients or organic load.

For water flows the Italian law sets out seven macrodescriptors: oxygen, ammonium nitrogen; nitrites, COD, BOD5, total phosphorous, Escherichia Coli. Moreover, for the determination of the ecological state a biological index (IBE) is also used.

The macrodescriptors for water bodies are those that define the trophic state, considering as limiting factors respectively phosphorous and nitrogen. The law therefore sets out the following macrodescriptors for water bodies: a-chlorophyll, transparency, total phosphorous, and hypolimnion oxygen. The environmental state of the ground waters is defined in the law on the basis of the quantitative state and the chemical state. For the evaluation of the latter, two sets of parameters are established: one that is compulsory and made up of hydrochemical parameters used to characterize the aquifer, and other additional parameters relative to dangerous pollution. The parameters are based on the following: electrical conductivity, chlorides, manganese, iron, nitrates, sulphates, and ammonium ions. The quantitative state measures the sustainable exploitation of the resource over a long period of time or rather the equilibrium between the withdrawal and the natural capacity of replacement.

Under the national legislation, in the model for the optimization of the water resources, both for surface water and for underground sources, the environmental state is summarized by an index which takes into consideration five possible values: (1) elevated; (2) good; (3) sufficient; (4) poor; (5) bad.

4.2.5 *Quality indices using hydrological scenario generation*

In a simplified frame for reservoirs, the problem of water quality can be studied considering the trophic state of water bodies strictly related to their artificial nature. Modeling trophic conditions of water bodies needs to take into consideration complex phenomena that are notably related to human activities in the basins. As is well known, a complete analysis of these phenomena needs to consider many relations deriving from chemical, physical, and biological aspects. In order to study the trophic state of a water body, the population density of the phytoplankton and of their limiting factors (sunlight, temperature, and nutrients) has to be analyzed. This requires a great effort to understand how these factors are related, after arranging them in a model using analytical relations. Of course even in a simplified approach the main characteristics of these relations is that they are not suitable for use in optimization modeling such as the WAMME project aims.

Literature presents us with various trials so that these aspects can be faced using mathematical modeling. Each one considers some factors in a way to simplify relations. Some modeling approaches consider relations involving many factors (Balzano et al., 1996; Gallerano et al., 1990) like phosphorous (sometimes also in different forms), nitrogen, other chemical elements (it depends on the specific case studied) that concern chemical factors, along with physical factors like temperature, wind stress, water inflows and outflows, and also some biological factors such as the main species of phytoplankton, in particular those that are the cause of eutrophication. Other simplified models, on the other hand, consider only some factors that can be directly related to the trophic state of the water body. Frequently the main variables are the phosphorous concentration (Vollenweider, 1968, 1975) sometimes coupled with nitrogen concentration (OCSE, 1982), the a-chlorophyll concentration (OCSE, 1982), the temperature, and others, singly or coupled (Reckhov, 1981). Obviously it is a great task to carry out a complete analysis of these phenomena in order to be able to implement and test the model that has been chosen.

Strictly considering a real water systems management optimization tool like WARGI, it is quite difficult to take into account these evolution aspects even if in a simplified manner. The first problem is that the time step needed to represent these phenomena is much smaller than the time step used in the optimization models. A significant analysis of the phenomena involved needs to first study the hydrodynamic fields and the temperature fields and second the concentration fields, the development of which needs very small time steps of integration (usually a few seconds). Another difficulty is related to an LP optimization approach, since they are nonlinear relations involving water quality evolution. Nevertheless, we tried to introduce in a very simplified manner a way to consider the water quality characterization in an LP water system optimization. The approach allows researchers to make predictions defining a synthetic index that explains the trophic state of the water body and allows the definition of a cost associated with the use of the water.

This index is related to the macrodescriptors introduced under the recent Italian legislation. In particular, concerning water bodies, reference is made to the a-chlorophyll concentration, which is a reliable index of the biomass present in the water body.

Preliminary and antecedent studies were made to consider these aspects (Carboni et al., 1998) using an index given by the rate of the a-chlorophyll concentration on the maximum a-chlorophyll concentration admitted for defined uses. To make previsions for the a-chlorophyll concentration, relations involving the stored volume of the basin have been used. This approach leads to a quadratic programming model. In the present version of WARGI another type of modeling has been tested. In particular the primary consideration that has been made is that the trophic state of a natural or artificial water body should depend on the hydrologic contribution. The data analysis shows that it is possible to consider a periodicity in the a-chlorophyll concentration trend, so the necessity to have linear relations suggests the treatment of the chlorophyll data by multiple linear regression analysis.

The South Sardinian lakes under study are continually checked by a specific EAF (Ente Autonomo del Flumendosa, the Regional Water Authority), so a lot of chemical, physical, meteorological, and biological information are available. The following data have been considered in WARGI optimization tool:

- Hydrologic inflows in the reservoirs
- Maximum a-chlorophyll concentration
- Minimum a-chlorophyll concentrations
- Average a-chlorophyll concentrations

Preliminary investigations considered the following time periods' aggregation: one, two, three, and six months. So, in each period we consider:

- The hydrologic inflow is the total volume flowed into the reservoir
- The maximum a-chlorophyll concentration is the maximum of the observed data in the period
- The minimum a-chlorophyll concentration is the minimum of the observed data in the period
- The average a-chlorophyll concentration is the average of the observed data in the period

For each water body and for each period the data at four depths has been considered: at the surface, at 1 m depth, at 2.5 m depth, and at 5 m depth; this was in respect to the fact that the most important growth of the phytoplankton is in the so-called photic zone (the zone interested by sunlight).

In the first phase, the regression analysis has been made considering the hydrologic contribution in the present and in the antecedent period, and the a-chlorophyll concentration (maximum, minimum, and average) at different depths.

The observed data are from four years, so at maximum we obtained 48 data for the monthly period, 24 data for the two-month period, 16 data for the three-month period, and 8 data for the six-month period.

In a second phase of data analysis a relation between a-chlorophyll concentration and time was considered. The trend of the observed data show that two peaks exist during the year and that these two peaks are separated by about six months. This relation has been developed exactly between observed datum and the date of the observation transformed by means of a cycling function of the time. To measure time from six months before the date of the peak, a sinusoidal function can be used to define a "transformed date."

The function used has this expression:

$$D_{transformed} = \sin\left[\frac{1}{2}(t - t_0) * \frac{2\pi}{T}\right]$$

where: $D_{transformed}$ is the date transformed by means of the function and referred to the selected origin; t is the date of the observation; t_0 is the date of selected origin; T is the time period in a-chlorophyll concentration (one year). The a-chlorophyll concentrations have been averaged in each period and dates transformed as said before. In the following analysis the averaged a-chlorophyll data have been considered as centered in each period.

In this preliminary work four reservoirs were considered. All these belong to Flumendosa-Campidano-Cixerri hydraulic system situated in the south of the Sardinia island and managed by EAF, the Regional Water Authority. The main characteristics of reservoirs are reported in Table 4.1. The reservoirs are: Cixerri reservoir; Mulargia reservoir; Flumendosa reservoir at Nuraghe Arrubiu section; and Simbirizzi reservoir. For each reservoir the total hydrological inflow has been considered in each period. Only for the Simbirizzi reservoir has it been necessary to make a balance to obtain the contribution of the reservoir. In fact, this is a small basin used as a storage reservoir that does not have a significant natural hydrological contribution. Its main water contributions or withdrawals are regulated by the Flumendosa system.

The observed a-chlorophyll concentration data for the Cixerri, Mulargia, and Flumendosa reservoirs started in January 1994 and finished in December 1997. For the Simbirizzi reservoir they started in March 1992 and continued until November 1995.

4.2.6   *Correlation between a-chlorophyll concentration and hydrologic contribution*

For all examined reservoirs, this analysis shows that the higher correlation coefficient has been obtained for time periods of three and six months. This, of course, is in part due to the small number of the data set used to make the linear regression analysis, but it is also possible to consider that the trophic phenomena can be related to events developing during longer time periods.

To confirm this, with the exception of the Flumendosa reservoir, usually for monthly and two-month periods, the correlation coefficient between average a-chlorophyll concentration and hydrologic inflow in the antecedent period is higher than the correlation coefficient between average a-chlorophyll concentration and hydrologic inflow in the same period. For three-month and six-month periods the opposite is usually true.

It is possible that, with the exception of the Mulargia reservoir, the higher values for the correlation coefficient have been achieved for the one-meter depth data set. For the Mulargia reservoir the higher values have been achieved for 2.5- and 5-meter depths. This is in accordance with the fact that the phytoplankton grows in the first layers of the water body and that the Mulargia reservoir is deeper than the others and has a lower biomass load.

Still, the correlation coefficient trend shows that the antecedent conclusions are true only for the correlation between hydrologic contribution and average a-chlorophyll concentration, whereas for correlation between hydrologic contribution and maximum or minimum a-chlorophyll concentration it is not possible to locate a univocal trend. In the following analysis we chose to consider only the average a-chlorophyll concentration.

Table 4.1 Main Reservoir Data

	Cixerri	Mulargia	Flumendosa	Simbirizzi
Total catchment basin [km²]	426	1183.16	1004.51	8.50
Reservoir surface at maximum level [km²]	4.90	12.40	9.00	3.20
Quote at maximum level [m s.l.m.]	40.50	259.00	269.00	33.50
Quote at maximum regulation level [m s.l.m.]	39.00	258.00	267.00	32.50
Volume at maximum level [$m^3 \cdot 10^6$]	32.20	347.70	316.40	33.80
Volume at maximum regulation level [$m^3 \cdot 10^6$]	23.90	320.70	292.90	28.80
Maximum depth at maximum level [m]	19.00	94	119	16.50
Average depth at maximum level [m]	6.10	25.87	35.16	10.56
Average annual hydrological inflow [$m^3 \cdot 10^6$]	90.57	18.29	250.64	0.39
Average annual total phosphorus contribution [t/year] (*)	25	10.8	10.9	1.4
Average annual total nitrogen contribution [t/year] (*)	Not available	217.4	342.6	22.8
Trophic state	Ipereutrophic	Mesotrophic	Oligotrophic-Mesotrophic	Ipereutrophic-eutrophic
Number of observations utilized	95	70	37	191
Main utilization	Agricultural Industrial	Urban Hydro-electrical Agricultural Industrial		Urban Agricultural

4.2.6.1　*Correlation between a-chlorophyll concentration and time*

This analysis shows a trend similar to the antecedent. However, for all reservoirs not including the Mulargia reservoir, the highest correlation coefficients have been obtained for one-meter depth. This is for all observed data sets and for the average a-chlorophyll concentration in each period. For the Mulargia reservoir the highest correlation coefficient values have been achieved for 2.5- and 5-meter depths.

4.2.6.2　*Multiple linear regression analysis*

The results reflect the preceding correlation analysis. The highest multiple correlation values have been obtained for the one-meter depth data set, with the exception of the Mulargia reservoir. Yet, for this depth the highest values have been obtained for three- and six-month periods. This is, in part, to be explained by the low number of data used to make the multiple regression.

For each basin, the multiple regression equation can be written:

$$chla_{calculated} = a\ h_1 + b\ h_2 + c\ h_3 + d$$

where: $chla_{calculated}$ obtained a-chlorophyll value by the multiple linear regression equation; h_1 hydrologic contribution in the same period; h_2 hydrologic contribution in the antecedent period; h_3 transformed time data as illustrated before; and a, b, c, d are coefficients of the equation.

The comparison between the observed average a-chlorophyll values and those obtained by the antecedent relations remark that it is only for six-month periods that there is usually a sufficient accordance between values, this in particular for the Mulargia reservoir. Instead, for monthly and three-month periods only in a few cases is it possible to see an acceptable correspondence.

4.2.7　*Evaluation of the potential use index*

In respect to the optimization algorithm the final step is the evaluation of an index to define the water quality state related to the final use. As made in an antecedent study considering the water quality in an optimization approach (Carboni et al., 1998), it is defined by the rate between the observed or calculated average a-chlorophyll value and the maximum value admitted for each use. Considering south Sardinian lakes, in many cases the calculated value of this index (as well as the observed one) is higher than the admitted.

The effort to find an easy way to consider the quality aspects of a water resource in a linear programming optimizations technique shows that it is difficult to take into account all these phenomena by means of simple linear regression analysis. This is due in part to the very small amplitude of the data set that we have, but mainly for the intrinsic complexity of these phenomena, which cannot be constricted in this simplified formulation.

It is possible to say that only for large time periods of data aggregation is it possible to find a satisfying correspondence between the observed values and those generated by the multiple regressive model. This of course is in part due to the limited data set. Nonetheless, it is possible to think that for these phenomena there will be a valid relation between the trophic state with the hydrologic contribution in a large time period. Of course it is necessary to test this method with higher numbers and more complete data sets.

For eutrophic reservoirs the proposed approach can be taken as a preliminary way for considering these water quality aspects inside an optimization model. In the same way we considered the a-chlorophyll concentrations. A more significant index related to the most important species that generate the biomass in each reservoir could be considered. We are on an index obtained from the rate between some algae species present in the studied water body. The general purpose is to obtain an index that is more strictly referred to each single reservoir (Marchetti, 1993).

4.3 The optimization package WARGI: Water Resources System Optimization Aided by Graphical Interface

As previously highlighted, the main aspect of improving the practical utility of optimization in water resources revolves around the possibility of the user to employ efficient computational codes to deal with large-scale problems, adherent to reality, and relying on an interactive graphical user interface (GUI) managing data. Probably, at the moment, this is the primary requirement for convincing decision makers to accept optimization as a real problem- solving tool in a DSS framework.

Consequently, when building WARGI, adequate consideration was placed on providing user-friendly optimization software for water systems applications. Following the preceding aims, WARGI has been developed as a general-purpose optimization tool characterized by considering a user-friendly interface, developed in a scenario-optimization modeling framework along with water quality indices. This enables the consideration of the possibility of conventional and marginal water utilization, which would definitely solve large-scale water system optimization problems.

The main features of the WARGI optimization tool are:

- Friendly to users in the input phase and in processing output results
- Prevents obsolescence of the optimizer exploiting the standard input format in optimization codes
- Easy to modify system configuration and related data to perform sensitivity analysis and to process data uncertainty

"Preventing obsolescence" and "easy updating" are strictly connected aspects. To prevent the risk of an early uselessness, the tool has been assembled as a transparent boxes collection, consisting of independent modules; in such a way each module can be easily managed.

The main boxes inside the graphical interface are:

- System elements characterization
- Topology connections and transfer constraints
- Links to hydrological data and demand requirements files
- Time period definition and scenario settlement
- Water quality indices attribution to sources, demands, and transfer elements
- Planning and management rules definition
- Benefits and costs attribution
- Call to optimizers
- Output processing

WARGI allows the start of the analysis from the physical system so that an optimal solution can be reached having the possibility of controlling all the intermediate phases. WARGI allows for an easy updating of the system configuration and considers different system optimizers using standard data-input format. The interface has been developed and tested within an HP-Unix and PC-Linux environment. The various software components have been coded in C++ and TCL-TK graphic language.

4.3.1   *Problem formalization*

Even if it is quite impossible to define a general mathematical model formalization for water resources planning and management problem, WARGI allows the consideration of the components of a system to be as general as possible based on the most typical characterization of this type of models. Different components can be considered or ignored updating constraints and objective. In the following we refer to the dynamic network $G = (N, A)$ where N is the set of nodes and A is the set of arcs. T represents the set of time steps t.

Following the physical-system formalization adopted in previous works (Sechi and Zuddas, 1997, 1998, 2000) and recently used in the EU-WARSYP project, the water resources system can be viewed as a physical network where nodes and arcs are as follows:

- Reservoir nodes: represent surface water resources with storage capacity. In these nodes losses by evaporation can be considered.
- Demand nodes: such as for civil and industrial irrigation among others. They can be consumptive or totally nonconsumptive water demand nodes.
- Hydroelectric nodes: nonconsumptive nodes with hydroelectric units.
- Confluence nodes: such as river confluence, withdraw connections for demands satisfaction, etc.

- Ground water nodes: represent ground water resources with storage capacity.
- Desalinization-plant nodes: represent the possibility of treating salt water.
- Wastewater-treatment plant nodes: represent the wastewater treatments for reuse.
- Treatment plant: represents water treatment for its use.
- Natural stream arcs: represent the natural runoff along rivers or riverbeds.
- Conveyance work arcs: artificial channels such as ditches, pipes, etc.
- Water pumping facility arcs: arcs with a pumping plant.
- Emergency transfers arcs: allow transfer of water to face shortage.
- Spilling arcs: allow direct injection of surface water from a connection node into an aquifer.

The operational management issues to consider in the problem can be easily modeled using graph structures such as the following:

- Priorities in the stored water level of reservoirs
- Priorities in the demand satisfaction of demand nodes
- Penalty on shortages and emergency transfers
- Water quality aspects related to water bodies conditions and demands requirements

The planning issues refer to the design of the physical system such as dimensions associated with future works: reservoir capacities, pipes dimensions, irrigation areas, etc. Other planning aspects are related to unit consumptive use demands, irrigation technologies, and agricultural assessments. Requested information can be given as scalar (constant in any period), cyclic (assuming the same values in homologue periods), or vector (varying in each period) values.

For a reservoir in an operational state (constructed reservoirs) the following information is required:

- Capacity: max storage volume for inter-periods transfer
- Ratio between max volume usable in each period and the reservoir capacity
- Ratio between min stored volume in each period and reservoir capacity
- Gradient of the relationship between the reservoir surfaces and volumes
- Evaporation losses per unit of reservoir surface
- Hydrological input to the reservoir
- Reservoir stored water quality index
- Hydrological input quality index
- Spilling cost

For a reservoir in a project state (reservoir to be constructed) the further following information is required: max allowed capacity; min allowed capacity; and construction costs. For a civil demand in an operational state the following information is required: population; unitary demand; request program; minimum required quality index; and deficit cost. For a civil demand in a project state the following further information is required: max population; min population; and net construction benefits. For an industrial demand in an operational state the following information is required: industrial center dimension; unitary demand; request program; minimum required quality index; and deficit cost. For an industrial demand in a project state the following further information is required: max dimension; min dimension; and net construction benefits.

For irrigation demand in an operational state the following information is required: agriculture center dimension; unitary demand; request program; minimum required quality index; and deficit cost. For an irrigation demand in a project state the following further information is required: max area to irrigate; min area to irrigate; and net construction benefits.

For a hydroelectric power station in an operational state the following information is required: production capacity; production program; energy efficiency; and production benefit. For a hydroelectric power station in a project state the following further information is required: max production capacity; min production capacity; and construction cost. For a confluence node the following information is required: hydrologic input (if arcs are natural streams) and input quality index.

For a ground water node the following information is required: aquifer capacity; ratio between the max usable volume and the aquifer capacity; ratio between the min stored volume and the aquifer capacity, ground water recharge; ground water quality index; input quality index; and spilling cost. For a pump station in an operational state the following information is required: pumping capacity; pumping program; pumping efficiency; and pumping cost. For a pump station in a project state the following further information is required: max pumping capacity; min pumping capacity; and construction cost.

For a desalinization plant in an operational state the following information is required: desalinization water production; production program; treated water quality; and desalinization cost. For a desalinization plant in a project state the following further information is required: max production capacity; min production capacity; and construction cost.

For a wastewater treatment plant in an operational state the following information is required: treatment capacity; treatment program; treatment quota; treated water quality index; and treatment cost. For a wastewater treatment plant in a project state the following further information is required: max treatment capacity; min treatment capacity; and construction cost. For a water treatment plant in an operational state the following information is required: treatment capacity; treatment program; treated water quality index; and treatment cost. For a water treatment plant in a project state the following further information is required: max treatment capacity; min treatment capacity; and construction cost.

For a transfer arc in operational state the following information is required: transfer capacity; ratio between max transferred volumes and capacity; ratio between min transferred volumes and capacity; assured quality index; and operating cost. For a transfer arc in a project state the following further information is required: max transfer capacity; min transfer capacity; and construction cost.

Consequently, variables considered in the optimization model can be divided into flow variables (or operational variables) and project variables. The flow variables can refer to different types of water transfer such as:

- Water transfer in space along arc connecting different nodes at the same time
- Water transfer in arc connecting homologous nodes at different times
- Water transfer in arc connecting nodes to the root node
- Water-losses arcs (as for evaporation) to transfer water to the root node
- Water transfer from the root node to demand nodes to face request in drought periods (deficit transfers)

The planning variables refer to the project state, and they are associated with the dimension of future works: reservoir capacities, pipe dimensions, irrigation areas, etc.

Constraint equations of the optimization model can be divided into the following types:

- Continuity equations for confluence nodes
- Continuity equations for reservoirs and aquifers
- Continuity equations for demand nodes
- Continuity equations for plant nodes
- Continuity for the root node
- Requests evaluation for the centers of water consumption
- Evaporation evaluation at reservoirs
- Losses evaluation for aquifers
- Production evaluation for plants
- Relations between flow variables and planning works
- Upper and lower bounds on spatial water transfers related to dimensioned works
- Upper and lower bounds on spatial water transfers related to planning works.
- Upper and lower bounds on temporal transfers related to dimensioned works
- Upper and lower bounds on temporal transfers related to planning works
- Upper and lower bounds on project variables

- Quality constraint equations at confluence nodes
- Quality constraint equations at reservoirs and aquifers
- Quality constraint equations at demand nodes
- Quality constraint equations at plant nodes
- Scenario constraints for reservoirs and aquifers at branching time

The OF considers costs and benefits associated with flow and project variables and can be divided into the following terms:

- Construction and operational costs for transfer arcs
- Construction and operational costs for reservoirs
- Construction and operational costs for demand centers
- Construction and operational costs for plants
- Penalties for deficits, losses, and target faults
- Weights associated with scenario optimization

4.3.2 *Identification of network components and sets*

Nodes:

res set of reservoir nodes: these represent surface water resources with storage capacity. In these nodes losses by evaporation can be considered.

dem set of demand nodes: such as for urban and industrial irrigation among others. They can be consumptive or nonconsumptive water demand nodes.

hyd set of hydroelectric nodes: they are nonconsumptive nodes associated with hydroelectric plants.

pum set of pump nodes: these represent pumping plants.

con set of confluence nodes: such as river confluence, withdraw connections for demands satisfaction, etc.

aqf set of aquifer nodes: these nodes represent ground water resources with storage capacity.

dsl set of desalinization-plant nodes: these nodes represent saltwater treating plants.

wtp set of wastewater-treatment plant nodes: these represent the wastewater treatments for reuse.

tpn set of treatment-plant nodes: these represent water treatment plants for its use.

Other sets of nodes can represent different types of water plants.

Arcs:

NAT set of natural stream arcs: these represent the natural runoff along rivers or riverbeds.

CON set of conveyance work arcs: these represent artificial channels such as ditches, pipes, etc.

PUM set of water pumping facility arcs: these represent transfers with pumping plants activities.

EMT set of emergency transfers arcs: these arcs allow transfer of water to face shortage.

SPL set of spilling arcs: these allow the overflow from reservoirs.

REC set of recharge arcs: these allow direct injection of surface water from a connection node into an aquifer.

LOS set of water losses arcs as for evaporation from lakes, ground water deep infiltration, etc.

4.3.3 *Required data*

Some of the operational management issues to be considered in the optimization problem can be easily modeled directly on the graph structures such as priorities in the stored water level of reservoirs, priorities in the demand satisfaction of demand nodes, and so on.

The planning issues refer to the design of the physical system such as dimensions associated with future works: reservoir capacities, pipe dimensions, irrigation areas, etc. Other planning aspects are related to unit consumptive use demands, irrigation technologies, and agricultural assessments.

Requested data can be given as scalar (constant in any period), cyclic (assuming the same values in homologue periods), or vector (varying in each period) values. Data marked with (+) are required for an operational state (existing works with a known dimension) while data marked with (*) are required for a project state (works to be constructed). No marked data are required for an operational and project state.

Required data for a reservoir $j \in$ **res**:

$Y_j(+)$	max storage volume for interperiods transfers
r^t_{jmax}	ratio between max volume usable in each period t and the reservoir capacity
r^t_{jmin}	ratio between min stored volume in each period t and reservoir capacity
d_j	gradient of the relationship between the reservoir surfaces and volumes
l^t_j	evaporation losses at time t per unit of reservoir surface area
inp^t_j	hydrological input in each period t to the reservoir
Cs^t_j	spilling cost in each period t
Cp^t_j	interperiod ($t \rightarrow t + 1$) transfer benefit for stored water
Qy^t_j	reservoir-stored water quality index at time t
Qh^t_j	hydrological input quality index at time t
Y_{jmax}	(*) max allowed capacity
Y_{jmin}	(*) min allowed capacity
g_j	(*) construction costs

Required data for a demand $j \in$ **dem**:

P_j	(+) center dimension
d^t_j	unitary demand
b^t_j	request program
c^t_j	deficit cost
Qp^t_j	minimum required quality index
P_{jmax}	(*) max allowed center dimension
P_{jmin}	(*) min allowed center dimension
b_j	(*) net construction benefits

Required data for a hydroelectric power station $j \in$ **hyd**:

H_j	(+) production capacity
b^t_j	production program
b_j	production benefit
H_{jmax}	(*) max production capacity
H_{jmin}	(*) min production capacity
g_j	(*) construction cost

Required data for an aquifer node $j \in$ **aqf**:

A_j	aquifer capacity
r^t_{jmax}	ratio between the max usable volume and the aquifer capacity
r^t_{jmin}	ratio between the min stored volume and the aquifer capacity
L^t_j	deep infiltration losses
R^t_j	ground water recharge
Cl^t_j	loss cost in each period t
Cp^t_j	interperiod ($t \rightarrow t + 1$) transfer benefit for aquifer-stored water
Qa^t_j	aquifer-stored water quality index at time t
Qr^t_j	ground water recharge quality index at time t

Required data for a pump station n $j \in$ **pum**:

C_j	(+) pumping capacity
b^t_j	pumping program
e_j	pumping efficiency
c^t_j	pumping cost
C_{jmax}	(*) max production capacity
C_{jmin}	(*) min production capacity
g_j	(*) construction cost

Required data for a desalinization plant $j \in$ **dsl**:

D_j	(+) desalinization production capacity
b^t_j	production program

c^t_j desalinization cost
Qd^t_j treated water quality index
D_{jmax} (*) max production capacity
D_{jmin} (*) min production capacity
γ_j (*) construction cost

Required data for a wastewater treatment plant j∈ **wtp**:

W_j (+) treatment capacity
b^t_j treatment program
s^t_j treatment quota
c^t_j treatment cost
Qw^t_j treated water quality index
W_{jmax} (*) max treatment capacity
W_{jmin} (*) min treatment capacity
γ_j (*) construction cost

Required data for a water treatment plant j∈ **tpn**:

Q_j (+) treatment capacity
b^t_j treatment program
c^t_j treatment cost
Qq^t_j treated water quality index
Q_{jmax} (*) max treatment capacity
Q_{jmin} (*) min treatment capacity
c_j (*) construction cost

Required data for a confluence node j∈ **con**:

inp^t_j hydrologic input (if arcs are natural streams)
Qc^t_j hydrological input water quality index

Required data for a generic transfer arc a∈ **TRF** ≡ ∪(**NAT, CON, PUM, EMT, SPL, LOS**):

F_a (+) transfer capacity
r^t_{amax} ratio between max transferred volumes and capacity in each
                 period t
r^t_{amin} ratio between min transferred volumes and capacity in each
                 period t
c^t_a transfer cost in each period t
Qf^t_j assured water quality index for transferred water
F_{amax} (*) max transfer capacity
F_{amin} (*) min transfer capacity
g_a (*) construction cost

4.3.4 *Constraints formalization in the optimization model*

As previously described, variables considered in the optimization model can be divided into flow and project variables. Flow variables refer to different types of water transfers in the multiperiod network: both in space (water transfer along arcs connecting different nodes at the same time), and in time (water transfer in arcs connecting homologous nodes, with storage capacity, at different times), as well as along "dummy arcs" (i.e., water transfer to face deficit at demand nodes). Project variables refer to works in a project state, and they are associated with the dimension of future works: reservoir capacities, pipe dimensions, irrigation areas, etc.

Constraints in the optimization model are used to represent a large variety of links and limitations in the system activities, as for example, mass balance equations concerning flow variables, demand requirements constraints concerning flow and planning variables, transfer constraints along arcs concerning flow and planning variables, losses (such as reservoir evaporation) evaluation and transfer to the root node, relations between flow variables and planning works, upper and lower bounds on flow and planning variables, etc.

In order to deal with large-scale problems, the actual version of WARGI has been specifically developed to treat LP and QP models.

For some system elements, related variables and corresponding constraints are described hereafter in more detail.

Considering reservoir nodes, we can refer to Figure 4.4 to illustrate the adopted WARGI schematization and related variables and constraints: y^t_j stored water at reservoir j at the end of period t that can be used in the next periods. These are flow variables, and they can be regarded as the water flowing along the interperiods arcs ($t \rightarrow t + 1$) connecting the homologous

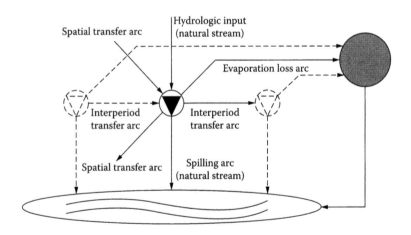

Figure 4.4 Reservoir node schematization.

nodes representing the reservoir j in the adjacent periods. The corresponding storage constraints, for each time period t are:

$$r^t{}_{kmin}Y_j \le y^t_j \le r^t_{jmax} Y_j \qquad j \in \textbf{res}$$

These constraints ensure that, in each period, the water stored (and transferred to a subsequent period) in reservoir j be in the prescribed range of allowed capacity. In an operational state Y_j are data that while in a project state are decision variables. In the last case they are bound by:

$$Y_{jmin} \le Y_j \le Y_{jmax} \qquad j \in \textbf{res}$$

where Y_{jmax} and Y_{jmin} are the bounds for the project variable Y_j.

Evaporation losses e^t_j from the reservoir can be evaluated from flow variables y^t_j as:

$$e^t_j = \delta_j \, l^t_j \, y^t_j \qquad j \in \textbf{res}$$

where the surface of reservoir j at the time t is evaluated using the δ_j ratio with the volume; the unitary evaporation from the surface is given by the l^t_j values.

The continuity (mass balance) equation for reservoir j can be expressed as:

$$y^t_j - y^{t-1}_j + inp^t_j + f^t_i - f^t_o - s^t_j - e^t_j = 0$$

where:

inp^t_j are the given data of hydrological input at time t to the reservoir j
f^t_i are flow variables as water coming from spatial-transfer arcs
f^t_o are flow variables as water released to spatial transfer arcs
s^t_j are flow variables representing spilling releases

Quality constraint for reservoir j ensures that water released or transferred from the reservoir meets the required indices. Implicitly we consider a linear behavior approximation using quality indices. The quality constraint for the reservoir can be written:

$$Qy^t_j \, y^t_j + Qf^t_i \, f^t_i + Qh^t_j \, inp^t_j \le Qy^{t+1}_j \, y^{t+1}_j + Qf^t_o \, f^t_o + Qy^t_j \, e^t_j + Qy^t_j \, s^t_j$$

Using the optimization tool WARGI a consistency analysis has been automatically carried out to verify the attribution of quality indices to system elements. Particularly, the reservoir outflows at the same time t must be characterized by same quality index, and this implies that $Qf^t_o = Qy^t_j$. To avoid nonadmissibility, the better (lower value) quality index for input water must be less in value than the assumed index for output water. The

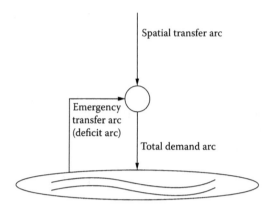

Figure 4.5 Demand node schematization.

WARGI consistency analysis for quality indices also considers values attributed to stored water in subsequent time periods.

Considering demand nodes, we can refer to Figure 4.5 to illustrate related variables and constraints: p^t_j water demand at civil demand center j in period t. These are flow variables associated with water transfer from the demand node to the root node. The corresponding constraint, for p^t_j evaluation at each time period t is:

$$p^t_j = \beta^t_j \, d^t_j \, P_j \qquad j \in \mathbf{dem}$$

where:

β^t_j assigned demand program
d^t_j unitary demand
P_j demand center dimension

These constraints ensure the fulfillment of the demand in each period, regardless of whether they come from the system or from a dummy resource (deficit arc). In an operational state P_j are data that while in a project state are project variables. In the last case they are bound by:

$$P_{jmin} \leq P_j \leq P_{jmax} \qquad j \in \mathbf{dem}$$

The continuity (mass balance) equation for civil demand node j can be expressed as:

$$p^t_j - f^t_j - x^t_j = 0$$

where:

x^t_j is the shortage at time t for demand j as deficit flow variable
f^t_j are flow variables such as water coming from spatial-transfer arcs

Quality constraint for demand-node *j* ensures that water transferred by spatial transfer arcs meets the required quality index Qp_j^t for the demand. The best possible quality index is implicitly attributed to the shortage arc; flows along this arc can consequently be worse than those required of the quality index in spatial water transfers coming to the demand center. The quality constraint for the civil demand node can be written:

$$Qf_j^t\ f_j^t + Qx_j^t\ x_j^t \leq Qp_j^t\ p_j^t$$

Using the optimization tool WARGI, a consistency analysis was automatically carried out to verify the attribution of quality indices to system elements. Variables and constraints are defined in the same way for other demand sets. Considering hydroelectric nodes, as they are nonconsumptive water demand nodes: h_j^t water through the plant *j* (hydroelectric power plant) in period *t*. These are flow variables associated to hydroelectric production. The corresponding constraints, for h_j^t evaluation at each time period *t* is:

$$h_j^t \leq \beta_j^t a_j^t\ H_j \qquad j \in \mathbf{hyd}$$

where:

β_j^t assigned production-program
α_j^t unitary-production water requirement
H_j production capacity

These constraints express the dependence of flow variables h_j^t on production capacity. In an operational state, H_j are data that, while in a project state, are project variables. In the last case they are bound by:

$$H_{jmin} \leq H_j \leq H_{jmax} \qquad j \in \mathbf{hyd}$$

No quality requirements are associated with hydroelectric nodes.

Considering a generic spatial-transfer arc, water flows on it must be bound to transfer capacity, determined by the dimension of the transfer work: f_a^t flow on arc *a* in period *t*. These are flow variables; the corresponding constraints for each time period *t* are:

$$r_{amin}^t\ F_a \leq f_a^t \leq r_{amax}^t\ F_a \qquad a \in \mathbf{TRF}$$

where:
ρ_{amin}^t ratio between admitted min flow and capacity
ρ_{amax}^t ratio between admitted max flow and capacity
F_a transfer capacity

These constraints ensure that, in each period, the transferred volume in arc *a* be in the prescribed range. In an operational state F_a are data that, while in a project state, are project variables. In the last case they are bound by:

$$F_{amin} \le F_a \le F_{amax} \qquad a \in \mathbf{TRF}$$

An assured quality index for water flowing is associated to these arcs. Quality constraints are directly associated to them. Nevertheless, using the optimization tool WARGI, a consistency analysis is carried out to verify the attribution of quality indices to arcs when considering the composition of system elements at nodes.

Variables and constraints are defined in the same way for other node and arc sets. Referring to the multiperiod dynamic network structure, mass balance constraints and quality constraints are defined in each node $i \in \mathbf{N}$. Moreover, lower and upper bound constraints may be defined in some arcs $a \in \mathbf{A}$ to represent some particular limits regarding transfer arcs in the set $\mathbf{TRF}$.

4.3.5 Objective function formalization in the optimization model

The objective function (OF) formalization used in WARGI can be considered as the sum of three parts: The first part concerns "weights" (costs, benefits, and penalties) on flow variables that, as previously described, are associated with flows in space, in time and along "dummy" arcs. Flow variables are related to hydrologic scenarios, and they can be "weighted" in different ways. Normally the weights are to be related to the probability associated with the scenario or otherwise, especially when considering equally probable scenarios, and they are also related to the "preference" or "critical nature" attribution of the scenario. A second part of the OF considers costs given to project variables and represents the "construction costs" on planned works. A third part of the OF is specifically required by the scenario optimization when the scenario optimization is used to define a "barycentrical" solution. Among homologous flow variables in different scenarios, a flow-variable distance minimization expression must be inserted in the OF.

Following these criteria and using the simplified notation previously defined, the objective function can be expressed as a minimization of the sum of the following three terms:

OF1:

$$w_g \left(\sum_{j \in res} \sum_{t=1,T} c_y^t y_j^t + \sum_{j \in dem} \sum_{t=1,T} c_p^t p_j^t + \sum_{j \in hyd} \sum_{t=1,T} c_h^t h_j^t \right.$$
$$+ \sum_{j \in pum} \sum_{t=1,T} c_b^t b_j^t + \sum_{j \in aqf} \sum_{t=1,T} c_a^t a_j^t + \sum_{j \in dsl} \sum_{t=1,T} c_d^t d_j^t c_w^t w_j^t$$
$$\left. + \sum_{j \in wtp} \sum_{t=1,T} + \sum_{j \in tpn} \sum_{t=1,T} c_q^t q_j^t \sum_{j \in TRF} \sum_{t=1,T} c_x^t x_j^t \right) = 1,G$$

OF2:

$$S_{j \in res} g_y Y_j + S_{j \in dem} g_p P_j + S_{j \in hyd} g_h H_j + S_{j \in pum} g_b B_j + S_{j \in dsl} g_d D_j + S_{j \in wtp} g_w W_j$$
$$+ S_{j \in tpn} g_q Q_j$$

OF3:

$$\Sigma_{g=1,G} \; w_g^* \; \Sigma_{s\in \text{TRF}} \; \| \; c_s \, (x_s{}^g - x_s^*) \; \|$$

We can refer to a compact standard form of the described LP mathematical model in order to focus attention on the problem of representing the uncertainty of the data series in an overall model. As usual, the compact deterministic model can be written:

$$\min \; c^T \, x$$

$$\text{s.t.} \; A \, x = b$$

$$l \leq x \leq u$$

where:

x	represents the vector comprehensive of all flow and projects variables
c	represents the cost vector comprehensive of all costs, weights, and penalties on operating and projects variables
l and u	represent vectors comprehensive of all lower and upper bounds on operating and projects variables
b	represents the vector of RHS
$A \, x = b$	represent the set of all constraints in standard form

In this way, the compact LP model includes data, variables, and constraints as described in preceding sections.

The scenario optimization model can be expressed as a "chance model" (Pallottino et al., 2002), describing the collection of one deterministic model for each scenario $g \in G$ plus a set of congruity constraints representing requirements of equal interstage flow transfers in all scenarios between two consecutive stages. Considering the system schematization given in Figure 4.6, the stochastic model will have the following structure:

$$\min \; S_g \, w_g \, c_g \, x_g \; + \; S_g \, w_g \, \| \; c_s \, (x_g - x_g^*) \; \|$$

$$\text{s.t.}$$

$$A_g \, x_g = b_g \qquad\qquad \forall_g \in \; G$$

$$x_g \geq 0$$

$$x_s \in \; S$$

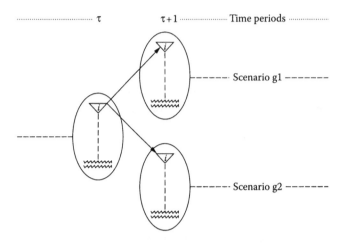

Figure 4.6 Segment of branch at branching time *t* referred to reservoir *j*.

where $x_s \in S$ represents the linking constraints on interstage flows, described in previous paragraphs, and x_g^* is the optimal policy ("barycentrical" flows) expected by the water manager. This kind of model can be solved by decomposition methods such as Benders decomposition techniques, which exploit the special structure of constraints. When the size of the problem becomes huge, it is possible to resort to parallel computing.

4.4   WARGI graphical interface

The WARGI graphical interface provides the possibility of inserting the basic system configuration components and connections between them. As shown in Figure 4.7, the main window presents an empty canvas, and the program window consists of several parts as described below. An extended illustration of WARGI features has been given in the WAMME project.

> *Title*: The name of the current graph (NoName if the graph is new and has not been saved).
> *Menu Bar*: The menu bar for selecting various options and tasks.
> *Scroll Bars*: Permit scrolling the canvas in order to use a canvas larger than the size visible on the screen.
> *Status Bar*: Multifunction bar providing information on graphic objects and also acts as a guide during graph construction.

The tool palette is a small window containing a set of graphical objects needed for creating graphs. In order to open the tool palette, the user has to pull up the View menu and select Drawing Tool. The tool palette window will open and can be dragged into a suitable position on the screen.

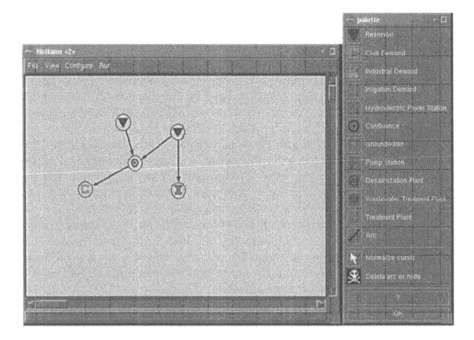

Figure 4.7 Main window of WARGI.

The tool palette consists of a number of buttons, each representing a graphic object or an action that can be performed on the palette or a graphic object placed on the canvas. Two buttons on the tool palette have an effect on the tool palette itself. The ? button toggles the text description on the right of the button icons. The text description is useful for new users who are not familiar with the symbols used for creating graphs. The OK button is used to close the tool palette once graph creation is complete.

Placing nodes on the canvas is a two-step process. First, the user must have the tool palette open. This will give the user access to the various node types that can be used. The first step to placing a node on the canvas is to *select the node* you wish to place from the tool palette. This is done by clicking on the icon representing the node in the tool palette using the left mouse button. The selected node type will be highlighted.

The second step is actually *placing the node* on the canvas. This can be done by simply moving the mouse over the canvas to the point in which the user wishes to place the node and then pressing the left mouse button. Once the node has been placed on the canvas the user may proceed to inserting the data associated with the node or continuing the graph-creating process.

Placing arcs is similar to placing nodes and requires three steps using the mouse. Using the tool palette it is possible to delete graphic objects from the canvas, selecting the Cancel Edge or Node from the tool palette. Once the graph has been completed or a new node or arc has been placed on the canvas it is necessary to insert the data related to the object. This is done through a template window, which is accessible by clicking on the graphic object using the right mouse button. The template window is different for each type of object placed on the canvas.

Some sections of the template windows can be accessible in the case of project optimization, others in system management (operational) optimization.

Once the graph has been completed and all the necessary data have been inserted, it is possible to generate the necessary MPS file to feed to the solver. MPS file generation is achieved by selecting Generate MPS from the File menu.

Graphs may be saved and previously saved graphs may be loaded by using the File menu. When a graph becomes too large to fit on the currently visible canvas, it may be helpful to scroll and zoom the canvas to have a better understanding and vision of the graph the user is creating. This can be done by selecting Scrolling and Zooming from the View menu.

In order to be able to launch the solver from the Graphical User Interface the user must first set the command line required to launch the solver. This is done from the Configure menu by selecting Solver. Once the required command has been inserted in the space provided clicking OK will set the command line for launching the solver.

At the end the user can visualize results using View results and Plot results options from the View menu.

4.5   Results applying WARGI to real cases

As extensively reported in the WAMME project, three WARGI applications to real water systems in the Mediterranean area were made: the Flumendosa–Campidano (Sardinia, Italy), Júcar (Spain), and Salso-Simeto (Sicily, Italy) water systems. Application criteria and results are summarized below.

4.5.1   Flumendosa–Campidano water system

Symbols meaning can be found in the palette-window reported in Figure 4.8. See Figure 4.9 for the template window for reservoirs. The schematization carried out using WARGI for the Flumendosa–Campidano system is reported in Figure 4.10. The water system consists of the following main elements:

- Ten reservoirs with a total capacity of 723 millions of cubic meters (Mm^3)
- Ten civil demand centers with a total request of 116 Mm^3 for year
- Nine irrigation demands with a total request of 224 Mm^3 for year
- Two industrial demands with a total request of 19 Mm^3 for year
- One hydroelectric demand with request of 90 Mm^3 for year
- Five pumping stations
- Nine water treatment plants
- Three wastewater treatment plants

Therefore, the total demand of the system was estimated as equal to 449 Mm^3 per year.

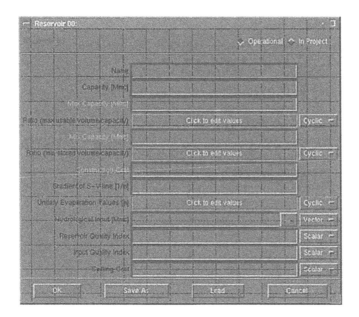

Figure 4.8 WARGI tool palette.

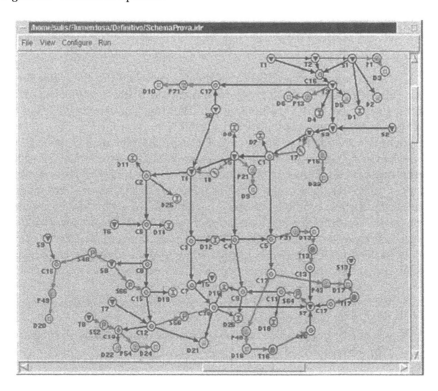

Figure 4.9 Example: The template window for reservoirs.

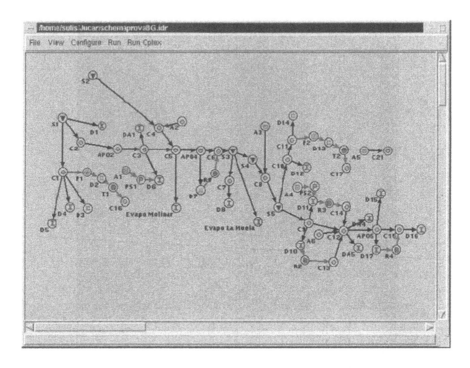

Figure 4.10 The WARGI input window for the Flumendosa–Campidano water system.

The water-quality classification of the reservoirs and of the water flows of the Flumendosa–Campidano system made use of the experience gained during the project and the availability of data that have been collected by the Flumendosa Water Authority since 1985.

Preliminary hydrologic analyses took into account the marked persistence of low runoff values in the past decade. Using WARGI, we first established the annual amount of volumes of the system to be issued considering the regional database (SISS) for the period 1922–1992. Further analysis was performed in relation to different synthetic scenarios of runoff in the rivers under consideration as the necessity of considering more hydrological scenarios was evident to characterize system performances. Nevertheless, to avoid an excess of complexity in the analysis, caused by the growth of examined sets, it was decided that investigations should be concentrated on considering two series:

A. Historical hydrological series: SISS (regional database) series 1922–1992 were considered;
B. Rescaled hydrological series: series prepared for the recent regional plan were adopted.

Thus, using the WARGI optimization tool, the results obtained using these two hydrologic input configurations were reported.

Moreover, the application of WARGI to the Flumendosa scheme has been carried out taking into consideration two phases with regard to quality constraint limiting water use:

1. During the first phase no quality constraints were given to limit the use of the resource on the basis of its quality;
2. During the second phase the indices of quality referring both to the resource and the water demand were considered.

Consequently, the optimization was carried out considering four situations given by the combination of two hydrologic configurations and two quality index attribution phases.

It was evident, however, from the analysis of the results, that maintaining the demand equal to that totally requested would have caused a deficit that was not admissible, particularly concerning the irrigation demand. This fact was highlighted especially by the results obtained with the use of the rescaled hydrologic series (B1 and B2 cases), which, on average, introduced a reduction of hydrologic input of about 50% of the average of the historical series.

Therefore, it was preferable to avoid the deficit reaching excessively high values. First, there was a reduction of the resource requested for hydroelectric use, reduced from 90 Mm³ to 20 Mm³, which was considered strategic for the production of energy in the region. Finally, indices of irrigation demand reduction were evaluated to annul the deficit in configuration B1. We, therefore, refer to cases B1b and B2b when reduction coefficients are applied to the optimization model, considering rescaled hydrological series without quality constraints (case B1b) or with quality constraints (case B2b). The same notation is adopted when we consider historical series (case A1b and A2b).

The introduction of the indices of quality concerning the resources of the system has caused modifications in the optimum configuration. Even if these modifications were not homogeneous throughout the system, the main impact on it can be observed considering the total amount of deficit in the historical and rescaled hydrological scenarios. Small deficits were highlighted in the Ab2 configuration. As can be expected, the rescaled hydrological (B2b) configuration shows much higher deficit values. To compare results, Table 4.2 and Table 4.3 report the synthetic resulting values obtained for demand deficits in the configuration A2b and B2b.

Table 4.2 Results Obtained for Demand Deficits in the Configuration A2b

Demand D	Years deficits >25%	Max year deficit [Mm³]	Year demand [Mm³]
All demands	0	14.36	257.92
Civil demand	0	2.88	93.32
Irrigation demand	0	6.96	125.6
Industrial demand	0	4.71	19.00
Hydroelectric Demand	0	2.45	20.00

Table 4.3 Results Obtained for Demand Deficits in the Configuration

Demand	Years deficits > 25%	Max year deficit [Mm³]	Year demand [Mm³]
All demands	16	153.13	257.92
Civil demand	1	24.75	93.32
Irrigation demand	20	112.67	125.60
Industrial demand	2	5.22	19.00
Hydroelectric demand	3	11.38	20.00

WARGI has shown that more than 30% of the overall deficit of the system was caused by the management criteria of the reservoirs and water release distribution to demand centers imposed by quality requirements.

4.5.2   Júcar water system

Basic information for this application was provided by the UPV partner (Universidad Politécnica de Valencia–Spain) in the WAMME project. UPV was also involved in the verification of results obtained using the WARGI model. The graphic scheme of the Júcar system is shown in Figure 4.11. The Júcar water system consists of the following main elements:

- Five reservoirs with a total capacity of 2085 Mm³
- Four civil demand centers with a total request of 173 Mm³

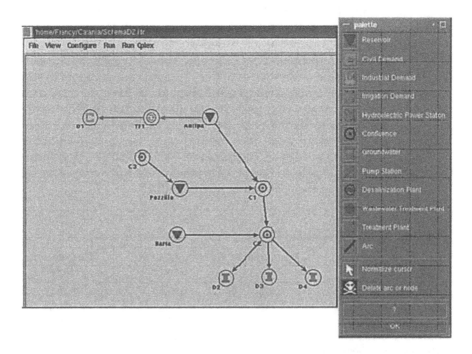

Figure 4.11 The Júcar water system in the second-phase WARGI model.

- One industrial demand center with a request of 35 Mm^3 per year
- Ten irrigation demand centers with a total request of 1188 Mm^3
- Five confluence nodes with hydrologic input
- Two pumping stations
- Two wastewater treatment plants
- Two water treatment plants
- Six ground water sources with a total capacity evaluated equal to 15,720 Mm^3

It should be pointed out that it is necessary to carry out some schematizations and simplifications in the system using the optimization tool WARGI when it is compared to the water system model, which uses the simulation approach. The main impact of this was evident when managing ground water sources. In particular, infiltration and outflow from the aquifer cannot be directly reproduced in the WARGI schematization. To partially overcome this difficulty, a two-stage WARGI modeling approach was used: the first one allows for getting flows in those arcs that are to be used to estimate ground water recharge, the second one to optimize the entire water resources system.

In a first modeling phase data on water quality are not given to the WARGI model. Quality indices are considered in a later phase.

Results obtained using WARGI, both considering or not quality constraints, ensured fulfillment of the water resources requirements of civil and industrial demand centers. No deficits were reported on these types of demand.

Irrigation demand nodes D5, D15, D16, and D17 had deficits resulting from optimization, and the critical period extends eighty to ninety years of historical hydrological series to be considered. To avoid the presence of concentrated irrigation deficits, each one of these irrigation centers was divided into three subcenters, giving each a scaled deficit cost. In Table 4.4 values characterizing deficits obtained in this configuration for demand centers are reported.

WARGI optimization can also be performed by inserting the quality indices on water resources and demands. Normally the introduction of quality indices constraints determines the growth of deficits caused by the

Table 4.4 Synthetic Deficit Values for Irrigation Center Without Quality Constraints

Demand	Years deficits >25%	Max year deficit (Mm^3)	Reliability (%)	Year demand (Mm^3)
All demands	0	110.23	67	1188
D5	9	27.71	80	80
D15	7	59.34	67	174
D16	9	24.82	67	79
D17	5	9.19	67	26

unavailability of water with acceptable quality conditions for demands. Nevertheless, as can be seen in the WAMME report, in the Júcar case, general conditions in behavior of quality seem quite stable: indices grow (this means less quality) from upstream to downstream, and users quality requirements show the same behavior. Therefore, civil demand remains completely satisfied, and only a slight growth occurs in irrigation demand deficits. The maximum year deficit only increases from 110.2 to 131.7 Mm^3 (the total irrigation demand is equal to 1188 Mm^3/year).

4.5.3   Salso–Simeto water system

Basic information for this application and verification of results obtained using the WARGI mode were provided by the DICA partner (University of Catania–Italy). The graphic scheme of the Salso–Simeto system is shown in Figure 4.12.

The Salso–Simeto water system consists of the following main elements:

- Two reservoirs (Ancipa and Pozzillo) with a total capacity of 151 Mm^3
- One diversion dam
- One civil demand center with a total request of 12 Mm^3 per year
- One irrigation demand with a total request of 121 Mm^3 per year
- One water treatment plant

In order to take into account the transfer of the winter spills from Ancipa reservoir to Pozzillo reservoir, two modeling steps have been performed: an

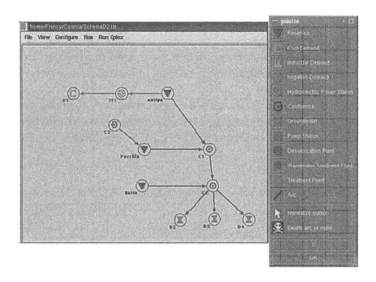

Figure 4.12 Final configuration of the Salso–Simeto system and WARGI tool palette.

Table 4.5 Performance Indices of the Salso–Simeto Water System Optimization

No. of violations to min. storage volumes (% months)		Sum of the annual squared irrigation deficits $(10^6 \cdot m^3)^2$	Maximum annual irrigation deficit $(10^6 \cdot m^3)$	Temporal irrigation reliability (%)	No. of years with deficit > 25% demand (–)
Ancipa	19.0	0	0	100	Municipal 0
Pozzillo	14.3	20,909	74.9	61.5	Irrigation 7

initial configuration has been set, in order to determine any water spills from Ancipa reservoir; and an updated configuration has been implemented, where the water transfers within the limit of the conduit maximum capacity are input to Pozzillo through the node C3, as reported in Figure 4.12.

To avoid the presence of concentrated irrigation deficits, the irrigation demand has been split into three equal parts: nodes D2, D3, and D4 in Figure 4.4, giving to each one a scaled deficit cost. Nevertheless, a higher priority of municipal water supply over irrigation is guaranteed. As a consequence of the deficit costs imposed for the municipal and the irrigation demands, there is no civil demand deficit during the optimization period, whereas irrigation deficits do mostly occur during the more critical historical drought periods. The performance of the system under the assumed conditions has been summarized in the computed indices that are shown in Table 4.5.

4.6   Conclusion

In the present work we illustrated a general-purpose optimization tool characterized by considering a user-friendly graphical interface. The WARGI package has been developed considering the objective of the WAMME project, which sets out to improve opportunities for using optimization, aided by a graphical user interface in water resource systems modeling. As previously highlighted, the package has been developed in a scenario-modeling framework and considers the possibility of conventional and marginal water utilization for solving large-scale water system optimization problems.

The usefulness of the developed optimization approach has been confirmed in the WARGI applications since the package allows the user an easy start to the optimization analysis by representing the physical sketch of the system in the main window of the toolkit. When requested information is given then the optimal solution can be reached. After this there is the possibility of viewing main results, modifying the configuration, and checking all the intermediate phases. WARGI allows for easy updating of the system configuration and considers different system optimizers using standard data-input format.

The benefit of resorting to mathematical programming, applied to optimization models solving real case water resources planning and management problems, seems to be confirmed in the multiphase DSS approach, as emphasized at the beginning of this chapter.

In the optimization phase we can solve a model representing the real physical problem in an obviously simplified manner, but which does not excessively reduce its level of adherence to reality. Consequently, in the simulation phase, we have only to examine reduced scenario configurations from the dimensional and temporal points of view. Moreover, if the optimization model remains adherent to the real problem, in a deterministic framework mathematical programming results give us the best management (in terms of water flows in the system) obtained by an ideal manager having previous knowledge of input and demands behavior in the system. These results can give us a measure of the "goodness" of the simulation-based DSS, at least for a reduced deterministic test-cases set.

References

Cannas, B., Fanni, A. Pintus, M., and Sechi, G. M. (2001a, July). *Alternative neural network models for the rainfall-runoff process.* Seventh International Conference on Engineering Applications of Neural Networks EANN 200, Cagliari.

Cannas B., Fanni, A. Pintus, M., and Sechi, G. M. (2001b). *River flow forecast for reservoir management through neural networks.* MODSIM 2001 Congress Proceedings, Canberra, Australia. Also published in the Proceedings of 28th Congress of Hydraulics and Hydraulic Engineering, Potenza, September 2002.

Cannas, B., Carboni, A., Fanni, A., and Sechi, G. M. (2000, July 17–19). *Locally recurrent neural networks for the water flow forecasting.* Sixth International Conference on Engineering Applications of Neural Networks. EANN 2000, Kingston Upon Thames, U.K.

CPLEX Optimization. (1993). *Using the CPLEX callable library and CPLEX mixed integer library.* Incline Village, Nevada.

Loucks, D. P., Stedinger, J. R., and Haith, D. A. (1981). *Water resource systems planning and analysis.* Englewood Cliffs, NJ: Prentice Hall.

Nicklow, J. W. (2000). Discrete-time optimal control for water resources engineering and management. *Water Int.* 25(1), 89–95.

Pallottino, S., Sechi, G.M., and Zuddas, P. (2002). *A DSS for water resources management under uncertainty.* IEMSs 2002 Conference on Integrated Assessment and Decision Support — IEMSs: International Environmental Modelling and Software Society, Lugano, Switzerland.

Rockafellar, R. T., and Wets, R. J. B. (1991). Scenarios and policy aggregation in optimization under uncertainty. *Mathematics of Operations Research* 16, 119–147.

Sechi, G. M., and Zuddas, P. (2000a). Scenario analysis in water resources system optimization under uncertainty conditions. Twenty-seventh Convegno di Idraulica e Costruzioni Idrauliche, (Vol. 3, 209–216). Genova. Also published in R. Mehrotra, B. Soni, K. K. S. Bhatia (Eds.), Integrated water resources management for sustainable development. Roorkee, India.

Sechi, G. M., and Zuddas, P. (2000b). WARGI: Water Resources System Optimization Aided by Graphical Interface. In W. R. Blain and C. A. Brebbia (Eds.), *Hydraulic engineering software*. WIT-PRESS, 109–120.

Sechi, G. M., and Zuddas, P. (2002). *The optimization package WARGI: Water Resources System Optimization Aided by Graphical Interface*. Proceedings of 28th Congress of Hydraulics and Hydraulic Engineering, Potenza, September 2002.

Simonovic, S. P. (2000). Tools for water management — A view of the future. *Water Int.* 25(1), 76–88.

chapter five

Decision support systems for drought management

Daniel P. Loucks
Cornell University

Contents

5.1 Introduction

About a quarter of the contiguous U.S. land surface (and about a third of the world's land surface) is semiarid or arid land. Water is a limiting resource in its development. Yet interestingly the most rapidly growing regions in the U.S. are states in the semiarid Southwest. The most rapidly growing countries in the world are concentrated in its semiarid regions. Engineering technology is providing the water from distant surface water supplies or ground water aquifers that fuels this development. Yet population pressures and pollution in these water scarce regions are causing overdrafts of both surface supplies and groundwater aquifers, making people more dependent on less reliable water supplies. All this coupled with the effects of climate are subjecting a growing percentage of the earth's population to increased risks of droughts and floods.

Droughts can be supply or demand driven. A shortage of water can result simply from lack of sufficient precipitation or excessive consumption. This shortage can be exacerbated by agricultural, municipal, and industrial water demands in excess of available water supplies. Recent droughts in regions spanning most of the world and their resulting economic, social, and environmental impacts underscore how vulnerable many of us are to this "natural" hazard.

Damages from droughts can exceed those resulting from any other natural hazard. In the U.S. the impacts of drought are estimated to average between $6 billion and $8 billion annually (National Drought Mitigation Center, 2003). Drought impacts occur primarily in agriculture, transportation, recreation and tourism, forestry, and energy sectors. Social and environmental impacts are also significant, although it is difficult to assign a monetary value to them.

Currently the Southwest portion of the U.S. is experiencing a 300-year drought. It is not yet clear what the total cost of this drought will be. Another severe drought period in the U.S. occurred over the years 1987–1989. Economic losses from that drought exceeded $39 billion (OTA, 1993; NOAA, 2002). This damage can be compared to the damages caused by the most costly flood, earthquake, and tropical storm events in the U.S.

The worst storm event in U.S. history was Hurricane Andrew. On August 24, 1992, this "costliest natural disaster," as it is called, hit south Florida and Louisiana. The storm killed 65 people and left some 200,000 others homeless. Approximately 600,000 homes and businesses were destroyed or severely impaired by the winds, waves, and rain from Andrew. Much of south Florida's communications and transportation infrastructures were significantly damaged. There was loss of power and utilities, water, sewage treatment, and other essentials, in some cases up to six months after the storm ended.

Andrew also damaged offshore oil facilities in the Gulf of Mexico. It toppled 13 platforms and 21 satellites, bent five platforms, and 23 satellites, damaged 104 other structures, and resulted in seven pollution incidents, two fires, and five drilling wells blown off location. The damage caused by Andrew in both south Florida and Louisiana totaled some $26 billion dollars.

The costliest earthquake in U.S. history was the Loma Prieta Earthquake. At five in the afternoon on October 17, 1989, the San Andreas fault system in northern California had its first major quake since 1906. Four minutes later, as over 62,000 fans filled Candlestick Park baseball stadium for the third game of the World Series and the San Francisco Bay Area commute moved into its heaviest flow, a Richter magnitude 7.1 earthquake struck. The Loma Prieta Quake was responsible for 62 deaths, 3,757 injuries, and damage to over 18,000 homes and 2,600 businesses. About 3,000 people were left homeless.

This 20-second earthquake, centered about 60 miles south of San Francisco, was felt as far away as San Diego and western Nevada. Damage and business interruption amounted to about $10 billion, with direct property damage estimated at $6.8 billion.

The most devastating flood in U.S. history occurred in the summer of 1993. All large Midwestern streams flooded including the Mississippi, Missouri, Kansas, Illinois, Des Moines, and Wisconsin rivers. The floods displaced over 70,000 people. Nearly 50,000 homes were damaged or destroyed, and 52 people died. Over 12,000 square miles of productive farmland were rendered useless. Damage was estimated between $15 and $20 billion. Again, the costs of the drought of 1988–1989 exceeded $39 billon.

The Andrew, Loma Prieta, and Mississippi events were sudden and dramatic. Droughts on the other hand are neither sudden nor dramatic. They are often not even given names other than their dates. Nevertheless, they can be much more costly. Drought planning and implementing mitigation measures can help reduce those costs.

5.2   Drought planning

Society's vulnerability to droughts is affected by population density and growth, especially in urban regions, and changes in water use trends, government policy, social behavior, economic conditions, and environmental and ecological objectives. Changes in all of these factors tend to increase the demand for water and hence increase society's vulnerability to droughts.

Although drought is a natural hazard, society can reduce its vulnerability and therefore lessen the risks associated with drought events. The impacts of droughts, like those of other natural hazards, can be reduced through planning and preparedness. Drought management clearly involves decision making under uncertain conditions. It is risk management. Planning ahead to identify effective ways of mitigating drought losses gives decision makers the chance to reduce both suffering and expense. Reacting without a plan to emergencies in a "crisis mode" during an actual drought generally decreases self-reliance and increases dependence on government services and donors.

5.3   Drought decision support

There are many aids to decision making. These aids include monitoring and forecasting facilities and capabilities, published rules of operation, flood and drought management plans with their triggers and special rules of operation, and a variety of planning, management, and real-time operating models. Each of these items supports decision making and thus could be called a decision support system. In this chapter the decision support systems I am referring to are interactive data-driven computer models built and used to estimate impacts of alternative water resources development and management decisions. These interactive data-driven models are also used to estimate the impacts of alternative assumptions about how particular water resource systems work and how they may work in the future given particular climatic, hydrologic, economic, and ecologic assumptions and scenarios — including drought conditions. Using such decision support systems stakeholders can build and run their own water resource system models and test their own assumptions regarding input data. In doing so they, the stakeholders, can reach a common or shared vision of at least how their physical system works and what is needed to make it work better — i.e., what must be done to meet both current and expected future needs and objectives.

Within the past year, the American Water Works Association Research Foundation has funded a three-year project to develop just such a computerized decision support system to aid utility strategic planners in effectively evaluating options for managing and developing reliable, adequate, and sustainable supplies of water for their customers "for the next 50 to 100 years." It is called "Decision Support System for Sustainable Water Supply Planning." With advice and assistance from several major water supply utilities in different regions of the U.S., the contractor (Tellus Institute, Boston, MA) is to develop a generic decision support system that will meet the long-term planning needs of any water supply utility. No small task! Meeting this goal will be a challenge in spite of all the experience many of us involved in this kind of work, including here in Valencia, have accumulated over the past several decades.

5.3.1 Background

The reduced quantity and quality of the available water and the increasing demands for water of high quality and at reasonable costs are of growing concern not only in the U.S. but also to many countries. As population increases over the new century, drinking water utilities will need to develop new sources, and customers will likely need to significantly change water use practices. Both supply and demand management will be needed. Developing new sources is becoming increasingly difficult due to competing agendas for water use from industry, agriculture, recreation, environmental concerns, and permitting requirements. The drinking water utility community in the U.S. is increasingly confronted with quantity issues, and the allocation of water rights is the source of constant and increasing debate. Simple procedural mistakes (e.g., not fully considering conservation before trying to develop a new source) can result in long delays. Many utilities currently need 5 to 10 years or more to get new sources permitted. Utilities are experiencing increasingly challenging permitting approvals for each incremental increase in the water supply. This emphasizes the need for advanced planning. As time passes, new source development is expected to be even more difficult.

Numerous approaches have been developed to help define the variety of social and physical ways that utilities can portray supply and demand effects on their watershed. Existing efforts have begun to go beyond water balances defined by the hydrologic cycle. The concept of a decision support system (DSS) for addressing water management issues at the watershed scale must consider ground water and surface water availability as well as the effects of water and land management measures on the functioning of eco-systems and public health.

Computer models have been developed, primarily in academia, which provide a basis for this comprehensive mass balance model. Some of the DSS models that have been developed provide the opportunity for incorporating a broader range of information (e.g., integrated resource planning, climate, in-stream flow, population, land use, etc.). Other models and DSS approaches include components such as importing geographical information system maps to define land use patterns. Land-satellite images are being used to evaluate the stress on large regions of wetlands resulting from overpumping ground water. Although an abundance of information appears available, and attempts have been made to pull together information, a comprehensive yet modular and easy-to-use tool as envisioned for this research effort has not been developed.

5.3.2 Planning DSS features

Developing this DSS will require a multidisciplined analysis, including insight into climate, hydrology, agriculture, ecology, recreation, population changes, urban planning, industry, economics, business management, and other pertinent models. Also fundamental to this project will be the consid-eration of drinking water utility needs in conjunction with other uses of the water supply. This includes both supply-side contributions to the water supply and demand-side water consumption components including both water delivered by the utility and other water consumption impacts to the water balance. In addition to this comprehensive water balance in a water-shed or river basin, a secondary objective in the development of the DSS is the consideration of a financial planning component.

The DSS is to include components that affect the water balance such as changing population, industry, agriculture, effects of climate, time and sur-face water quantity required for regeneration of aquifers, in-stream flow regimes necessary for maintaining diverse ecosystems, enhancing recreation and conservation, and in some cases hydropower and navigation (barge transportation). Supply-side options could include the identification of new surface or ground water supplies, increasing storage capacity, desalination, enhancing existing ground water supplies by conjunctive use or aquifer storage and recovery, and reuse (either indirect or as a substitute for potable supplies). Demand-side options could include big picture issues such as global warming resulting in additional evaporation. It could also include issues that may appear to have lesser impacts to the system such as distri-bution system leaks or conservation practices. Some components of the balance may be considered on both sides of the equation as long as the user does not double count a quantity of water for both reducing demand and increasing supply. For example, displacement of potable water use with alternative supplies such as use of cisterns or seawater to fight fires could reduce demand for potable water. Therefore, the DSS developers need to be specific about how the system will be defined.

The DSS must permit and facilitate sensitivity analyses of alternative assumptions and scenarios. The sensitivity analysis should allow portions of the balance to be held static and other components selectively altered to provide insight into the impact various efforts would have on the water supply. The water balance should allow for limits to be placed on certain components such as the size of a reservoir or minimum in-stream flow. Another part of this effort will be to consider components that can be entered into the water balance to show the impact of time variances.

Secondary to the comprehensive water balance, the researchers will identify financial planning components of the DSS. The financial planning components of the model will help utility planners evaluate the cost-effectiveness of developing a new water supply source (e.g., reservoir or new well field) or a new demand management option (e.g., low-volume toilet replacement program). Financial planning components may include the construction and operation, maintenance, and repair costs of new infra-structure projects to develop new water supplies. It should include the costs, as well as the benefits, of alternative demand management options and alternative water and wastewater pricing policies and rates.

The goal will be to design a system that will allow a broad range of inputs to the system, including inputs from models that utilities have developed for their own systems, yet also providing a user interface that will allow the DSS to be operated by people who do not have an extensive modeling background. The DSS model should be usable by city and utility strategic planners. Output from the model should be expressed in terms that are common to those professions as well as be comprehensible to the variety of stakeholders, each having their own specific information needs.

The DSS simulation model should allow varying time steps, say from daily to multiple year durations, in the same simulation, depending on how far into the future one is simulating. Daily increments may be needed in the short term, especially for operational studies, and for more strategic planning monthly and annual increments may be appropriate for near-term (i.e., up to 10 years) and mid-term (i.e., between 10 to 50 years) planning. Long-term planning might have 5- to 10-year windows. Adaptability of the DSS to advancing technologies should also be considered.

5.3.3   System calibration, verification, and testing

Components of the DSS must include routines that permit calibration of the values of physical parameters used to predict runoff, flow of water under the surface of a watershed, water quality, etc. Trial tests of the model are to be made in cooperation with several utilities using their site-specific geo-graphic and hydrogeologic information. Such tests will not only permit model refinement but also interface refinement and modification.

The DSS should provide insight to water management issues at the watershed scale and effects of water and land management. An example of this end-product may be a spreadsheet tool that models supply-and-demand data from a current baseline condition all the way through service area buildout. This tool should have the capacity to model supply and demand under different scenarios that could include different supply-and-demand management options. It should also track utility finance and capital expenditures as well as water and wastewater rates and charges. The goal of this tool should be to help utilities select between a range of supply and management options to help ensure a safe, reliable, and sustainable water supply for the community at buildout. Ultimately, the DSS tool will help planners to identify how utilities can develop new long-term supplies and avoid the pitfalls that hold up new supply development and permits for 5 to 10 years or more.

5.3.4 *The prototype model*

This new system will consist of two complementary and interactive parts — a knowledge portal and a water balance tool. The knowledge portal will be developed so that it can be used to develop analytical scenarios (e.g., data sets) that can interact with the water balance tool. The water balance tool (initially assumed to be the DSS model called WEAP) will be developed so that it can be used in conjunction with the knowledge portal or as a stand-alone software application for detailed water supply planning.

The knowledge portal will function as a central repository of analytical tools and relevant information for utility planners. The primary organizational structure will be thematic (e.g., climate change, water quality, ground water), although many items will span multiple themes. Each theme will be organized by categories of supporting materials. Categories might include tools, articles, case studies, data sources, contacts, and discussion forum.

The knowledge portal will be accessed via its own Internet website, enabling instant and universal access to its dynamic content by utility and strategic planners, as well as stakeholders. Information and data will either reside locally on the website or be linked to its original source on the web. All local information and links will be stored in a centralized database server, to facilitate searching, updating, and displaying of information, at minimal ongoing cost. Participants will be able to submit their own information, keeping the site up to date. A discussion server will foster interaction among participants and allow for annotations to be made to any information on the site. The dynamic and interactive nature of the knowledge portal is essential to its usefulness, far surpassing the worth of any static compendium.

The water balance tool will be the centerpiece of the DSS, helping planners evaluate a full range of future scenarios, potentially including assumptions on changing technologies, policies, demographics, economics, ecosystems, land use, and climate. Sensitivity analysis and scenario comparisons will facilitate the exploration of options and possibilities, costs and benefits.

The water balance tool will be comprehensive, incorporating the aspects relevant to sustainable water supply planning. The tool should be transparent and flexible so that the planners understand the underlying relationships embodied in the models and have the ability to modify them. Many components will be incorporated into the tool, such as water quality, conjunctive use, financial planning, ground water/surface water interaction, and hydrology, utilizing straightforward algorithms. Like a spreadsheet, the tool will allow the user to create new variables and express moderately complex functions and relationships. In cases where more complex algorithms are required, the tool will be able to dynamically and automatically link to external models (e.g., GCMs or various water quality models) through the knowledge portal. Planners should be able to use their preferred approaches rather than being forced to accept the results of a black box.

5.3.5   DSS use

The development of scenarios is at the heart of the decision support system, by providing planners with an understanding of the breadth of possible futures that may be faced and some knowledge of their likelihood through the use of sensitivity analyses. Over the course of the proposed 50- to 100-year planning horizon, a number of planning elements that may not be critical to short- and medium-term planning will likely take on added weight. Chief among these is the issue of climate change and sensitivity analysis based on a range of potential climate scenarios. Another element that has the potential to substantially impact long-term water supply planning is population forecasts. These are characterized by high levels of uncertainly and hence are candidates for sensitivity analysis. Another candidate for sensitivity analysis is the flow regime required to support ecosystems. The design of the DSS must accommodate and adapt to new information on the water required to meet ecosystem objectives.

5.4   Case examples

5.4.1   The Rio Grande watershed

The portion of the Rio Grande Basin that extends from its headwaters in Colorado into New Mexico is often arid. It also faces increasing demands for water resulting from population and economic growth and environmental water needs. It is likely, if not inevitable, that a severe drought will affect this region and cause significant economic damage. Coordinated management strategies are needed to deal with droughts that affect substantial portions of the Rio Grande watershed and that may affect the states of Texas, New Mexico, and Colorado (Ward et al., 2001).

To test whether new interstate institutions that coordinate surface water withdrawals and reservoir operations could reduce economic losses from droughts and to identify hydrologic and economic impacts of possible changes in management institutions that cope with droughts, a simulation model was developed to keep track of economic benefits, subject to hydrologic and institutional constraints. The modeling approach reflected the highly variable and stochastic supplies of the Rio Grande as well as fluctuating water demands. The model incorporated the hydrologic connection between ground water pumping and flows of the Rio Grande into the model. The Rio Grande Compact agreement of 1938 was built into the model to ensure that institutional constraints were met in the simulations.

Water supplies and flows in the watershed were represented in a yearly time-step over a 44-year planning horizon. Agricultural water uses were identified, including those in the El Paso Irrigation District. Municipal water demands in El Paso were represented. Total economic benefits were calculated for long-run normal inflows and a sequence of droughts, based on historical inflows from 1942 to 1985. Total drought damages were computed as the reduction in future economic benefits that would occur if flows dropped from average levels of 1.57 million acre-feet (MAF) (1936 million m^3) per year to 1.4 MAF (1726 million m^3) in drought years.

Three water development and management scenarios were evaluated: (1) increasing carryover storage at Elephant Butte Reservoir in New Mexico by reducing releases to downstream areas, (2) investments in irrigation efficiency in the Middle Rio Grande Conservancy District in New Mexico, and (3) constructing an additional 10,000 acre-feet (AF) (12.33 million m^3) of reservoir storage in northern New Mexico above Cochiti Lake.

Long-term annual average drought damages were estimated at $8 million for Texas, $5.8 million for Colorado, and $3.4 million for New Mexico (about $101 per acre-foot or 8 cents per m^3) reductions in water supplies. Increasing reservoir storage at Elephant Butte created a $433,000 annual loss for Texas and a $200,000 annual deficit for New Mexico. Improving irrigation efficiency in the Middle Rio Grande District resulted in a $15,000 annual benefit for Texas and a projected $7000 annual gain for New Mexico. The cost of implementing improved irrigation technologies would have to be very low to justify these investments economically. Creating additional reservoir storage at Cochiti Lake would create an annual benefit of $685,000 for Texas and an estimated gain of $134,000 per year for New Mexico.

This project demonstrates how optimization models can be utilized to evaluate the hydrologic and economic implications of multistate water management measures. The report suggests this type of model may be especially useful, if it can be expanded to include a mass surface water balance for the region, if it can better simulate groundwater pumping and return flows, and if it can include refined estimates of environmental needs and water use.

5.4.2   The Finger Lakes Region in New York State

Lake Cayuga is one of the so-called Finger Lakes in the Oswego River Basin. As shown in Figure 5.1 the Oswego River Basin is just south of, and drains

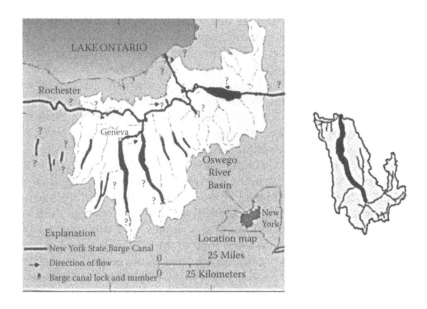

Figure 5.1 The Oswego River Basin and the watershed draining into Lake Cayuga in central New York State.

into, Lake Ontario, one of the Great Lakes. Lake Cayuga is one of the two largest Finger Lakes in the basin. The watershed draining into Lake Cayuga is being studied and managed by an interagency group. It has developed and is using a decision support system to help both better understand and manage their basin. This DSS is a generic simulation model capable of simulating any water resource system.

In this application the interagency personnel drew into the program's graphical interface the system configuration and watershed areas. They also entered the data that permit the program to perform a daily simulation of rainfall-runoff processes, streamflows, interactions with ground water, and of the transport from the land to the streams and eventually the lake of various water quality constituents, including sediment.

Figure 5.2 through Figure 5.5 illustrate part of the interface of the DSS, as applied to the Cayuga Lake Watershed. Although the interfaces may differ somewhat, many such DSSs have been constructed and are being used to assist water resource managers.

5.5 National drought management planning

The U.S. Army Corps of Engineers has had considerable experience using STELLA programs to develop and implement what they refer to as shared vision models (Werick, 2002; Werick and Whipple, 1994). They were used extensively during the national drought management planning studies in the U.S. about a decade ago (U.S. Army Corps of Engineers, 1991).

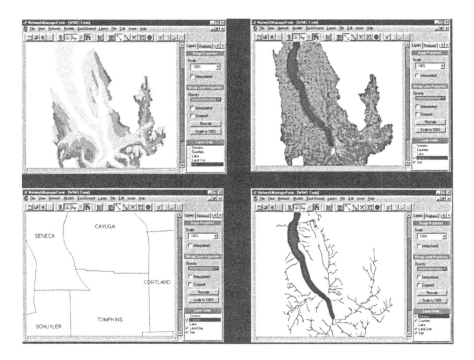

Figure 5.2 Data layers showing clockwise from upper left topography, land use, political boundaries, and streams draining into Lake Cayuga.

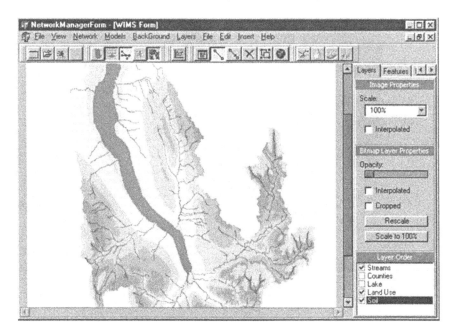

Figure 5.3 Transparent overlays of three of the data layers shown in Figure 5.2.

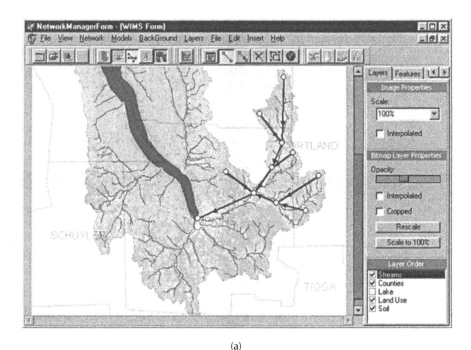

(a)

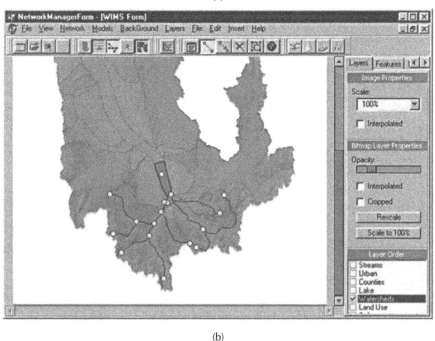

(b)

Figure 5.4 Ways of drawing in and displaying a network of streams, lake segments, and other surface water features such as gage sites, wastewater treatment discharge sites, monitoring sites, diversions, etc.

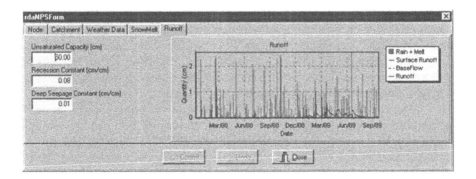

Figure 5.5 Time series output of precipitation (rain or snow melt), surface runoff, base flow, and total runoff. Meteorological and other data are available by clicking on the other tabs at the top of the window.

The DSSs used by the U.S. Army Corps of Engineers are typically relatively simple, not too data demanding, and can give a first cut at what may need more research and study and what may not, as water resource managers work to find management policies that best satisfy stakeholder objectives. In many regions, these objectives and expectations seem to be in a constant state of flux. As individuals learn more about what impacts what, and what people want from their water resource system, they usually need to alter their management policies and adapt to this new information. The processes of monitoring, analysis, increased understanding, and then action need to take place continuously. This succession of steps has been called adaptive management. It will be with us on into the future.

5.6 Conclusion

Planning for droughts is essential, but it may not come easily. There are many constraints to drought planning. For example, it is hard for politicians and the public to be concerned about a drought when they are coping with a flood — or any other more immediate crises. Unless there is a drought emergency, it is often hard to get support for drought planning. There are always more urgent needs for money and people's attention. Where coordination among multiple agencies can yield real benefits, it is not easy to get it to happen only when it needs to happen, e.g., during a severe drought. Multiagency cooperation and coordination must be planned for and practiced perhaps in virtual drought management exercises, in advance of the drought. Getting multiple agencies to work together only in a crisis mode is never efficient. Crisis-oriented drought response efforts have been largely ineffective, poorly coordinated, untimely, and inefficient in terms of the resources allocated.

Drought planning will vary from one city or region to another just because resources, institutions, and populations differ. Although drought contingency plans may vary in detail, they all should specify a sequence of increasingly stringent steps to either augment supplies or reduce demands as the drought becomes more severe — i.e., as the water shortage increases. This should happen in such a way as to minimize the adverse impacts of water shortages on public health, consumer activities, recreation, economic activity, and the environment in the most cost-effective manner possible.

Drought plans provide a consistent framework to prepare for and respond to drought events. A drought plan should include drought indicators, drought triggers, and drought responses. It should also include provisions for forecasting drought conditions, monitoring, and enforcement (IWR, 1994). Drought plans should consider a wide range of issues and be compatible with the political and social environments that can affect just what measures can be implemented.

The process of developing a drought plan and keeping it current is a continuing process that should include an informed public. Drought plans should also include measures to educate the public and keep them aware of the potential risks of droughts and measures that will be implemented to mitigate those risks. A comprehensive public information program should be implemented to achieve public acceptance of and compliance with the plan. Simultaneously, enforcement measures are necessary to encourage the public to abide by the water-use restrictions. Enforcement measures traditionally include penalties for noncompliance, but they can also include economic incentives such as rebates on low flow showerheads and faucets and cheaper water rate charges for lower consumption rates.

References

Institute for Water Resources (IWR), U.S. Army Corps of Engineers. (1994, September). *Managing water for drought.* IWR Report 94-NDS-8.

National Drought Mitigation Center. (2003). University of Nebraska, Lincoln, website. http://www.drought.unl.edu/.

OTA (Office of Technology Assessment). (1993). *Preparing for an uncertain climate,* vol. I. OTA–O–567. Washington, DC: U.S. Government Printing Office.

U.S. Army Corps of Engineers. (1991). *The national study of water management during drought, a research assessment.* Institute for Water Resources, IWR Report 91-NDS-3.

Ward, F., Young, R., Lacewell, R., King, J., Frasier, M., McGuckin, C., DuMars, C., Booker, J., Ellis, J., and Srinivasan, R. (2001, February). *Institutional adjustments for coping with prolonged and severe drought in the Rio Grande Basin.* New Mexico Water Resources Research Institute (NMWRRI) TR 317.

Werick, W. J. (2002). *Shared vision planning, a hyperlinked how-to guide.* Available at http://www.iwr.usace.army.mil/iwr/svtemplate/Introduction.htm.

Werick, W. J., and Whipple, W., Jr. (1994). *National study of water management during drought: Managing water for drought.* IWR Report 94-NDS-8. Alexandria, VA: U.S. Army Corps of Engineers, Water Resources Support Center, Institute for Water Resources.

chapter six

Methodology for the analysis of drought mitigation measures in water resource systems

Joaquín Andreu and A. Solera
Universidad Politécnica de Valencia, Spain

Contents

6.1　Introduction

This chapter deals with the analysis of measures applied to mitigate the effects of drought in developed water resource systems. What people normally understand by drought is really a series of phenomena related to the presence of water in the different phases of the hydrological cycle. Its first manifestation, and the origin of the whole process, is the "meteorological drought," which may be defined as a period of time during which precipitation remains below a certain threshold.

Within the hydrological cycle, precipitation is a signal that is transformed through the processes of evaporation, infiltration, storage in the earth, evapotranspiration, deep infiltration, both underground and surface storage and flows, surface runoff, etc. The repercussions of a meteorological drought are especially important in the moisture content of the ground, in the volume of rivers and springs, and in underground storage.

The repercussion of a meteorological drought on moisture content of the ground is particularly important due to the fact that many species, especially plants, depend solely on the water naturally available in the ground to survive and reproduce. A ground moisture drought or "edaphological drought" could be defined as that period of time during which the ground moisture content remains below a certain threshold.

The repercussion of a meteorological drought on the replenishing of the natural underground water tables (aquifers) and surface water (for example, lakes) and their subsequent outflows in the form of rivers and springs may cause a hydrological drought, which could be defined as that period of time during which the volume of water in rivers and springs remains below a certain threshold.

In all of the foregoing definitions, the threshold for the definition of the start of a drought is not necessarily the same at all times of the year, but could vary according to the season. It is quite frequent for this curve to be related to the curve of the average values of the respective variables used to define the different types of drought.

The study, description, and monitoring of these previously defined droughts has been developed over the course of many years (Wilhite and Glantz, 1985; Andreu, 1993; Buras, 2000; Loucks, 2000; Ito et al., 2001). The methods vary according to the type of drought under study and the aspect under consideration. On one hand, the probability approach tries to identify the statistical characteristics of the phenomena with the aim of obtaining data on distribution, intervals between droughts, and other results of interest. On the other hand, use is often made of indices to monitor different periods of drought. In addition, another dimension is added to the analysis, description, and monitoring of droughts when these procedures are carried out on a regional, instead of local, scale.

6.2   Operative drought

Unlike the droughts we have defined above, which are converted from one type to another through natural processes in the hydrological cycle, a developed water resource system is one in which the availability of water for diverse uses, including the ecosystem, does not depend only on natural processes, but also on processes controlled by man (Sánchez et al., 2001). In this way, unlike the previous cases, the same original signal could give rise to different results depending on how the artificial elements that compose the water resource system are managed and operated.

In the previous definitions of droughts the availability of water is analyzed, either in the form of rain or ground water or the water in rivers and springs, and if the quantity is below a certain threshold then we say there is a drought.

In the developed water resource systems, once the requirements of water for different uses and for the environment have been identified, if the available water resulting from natural sources and from the management and operation of the system does not meet these requirements, then it could be called an operative drought, in order to differentiate it from the previous types and to stress the importance of the operation of the system in the presentation and characteristics of this type of drought.

One often finds this type of drought referred to as socioeconomic (Vlachos and James, 1983), fundamentally because the shortage of water for the uses that depend on a water resource system produces financial losses and has social effects. However, other types of drought also produce these effects (for example, an edaphological drought also affects nonirrigated crops, as well as livestock pastures, forestry enterprises, etc.), so we do not think it appropriate to use this term to refer to operative droughts.

It could also be said that it is neither necessary nor appropriate to use the term *drought* to mean a failure in the water supply for different uses. But, since most of the time these failures are caused by natural droughts, we understand that the operative drought is the result of a natural drought in the system of water resources. In many highly developed basins, most of the effects of a natural drought are perceived as those of an operative drought.

Another consequence of an operative drought is the added environmental cost and the drop in water quality usually associated with droughts, which is frequently aggravated by waste discharges or by the reincorporation into the system of used water.

6.3   Time scales and the space factor in the analysis of operative droughts

Before continuing, we must draw attention to the fact that drought analysis gives different results for different scales of time and space. For an analysis to give relevant information for decision making, the choice of these scales is important.

In an arid or semiarid region, prolonged periods without rain are frequent (i.e., days or even months without precipitation). But, both the ecosystem and the agricultural and commercial activities in these regions have adapted themselves to these circumstances, so that to analyze a meteorological drought on a daily or weekly scale does not usually give useful information. The scale of the analysis must be at least monthly, and the most appropriate may even be yearly, depending on the type of drought and on the storage capacity of the system. But, as the annual scale is not suitable for recording of most of the hydrological phenomena that, as we shall see, it will be necessary to model, the monthly scale gives a compromise between the quantification of the results and the realistic recording of the phenomena.

In a developed water resource system an action at any point in the basin may have a direct or indirect influence at other points of the same basin, so that, apart from a few exceptions, the most appropriate spatial scale is that of the complete basin. The analysis of individual elements of the system or subsystems may give rise to erroneous conclusions due to the interdependence among the subsystems, both in resources (e.g., the relation between surface water and underground water) and in uses (e.g., return of used urban water capable of being reused). Therefore, it is essential to consider as a whole all sources of supply, water requirements, and any other elements that go into creating a system for the existing basin. It could even be necessary to analyze a space larger than a basin, if there were connections among different basins or if the supply for a certain use were to come from more than one basin.

Consequently, in the analyses carried out in the course of the work described in this chapter, the period of one month and the area of a complete basin were chosen as the default scales.

6.4 Analysis, characterization, and monitoring of operative droughts

Since the definition of an operative drought was given as a deficit with respect to certain necessities, the sequence of deficits is the basic information for the analysis of operative droughts. An operative drought event would therefore be a series of consecutive time units (e.g., months) in which there were deficits. An analysis of historic operative droughts can therefore be made similar to those carried out on other types of drought, based on the spells of drought, taking as variables of the analysis the duration, intensity, and the magnitude of these spells.

Also, for the exploitation phase of water resource systems, it is necessary to determine the situation at all times regarding the possibility of actually being in, or the prospect of soon being in, a situation of operative drought. Some of the indices used for this were Palmer's severity index (Palmer, 1965), the surface-water supply index, the scarcity index (U.S. Army Corps. of Engineers, 1966, 1975), the generalized scarcity index, and the index of the Sacramento River in California.

However, these analyses and monitoring of historical operative droughts do not provide information on the following points:

- The possibilities of the system experiencing future droughts: This is fundamentally due to the fact that the system and its future behavior will not be the same now as in the past, either in hydrology or in the established water uses and requirements, or in the available infra-structure and its management and operation.
- The effectiveness of possible mitigation measures: The above-mentioned analyses have only a descriptive utility, as do most of the indicators and characteristics of other types of drought, and they are unable to predict changes in the indicator as a result of using a certain mitiga-tion measure (except, of course, for simply defined measures with few implications for the rest of the water resource system).

It therefore becomes necessary to have available, as well as the above-mentioned indicators (or others that will be mentioned later), some kind of tool that will enable us to evaluate the possibility of future droughts and the effectiveness of mitigation measures against operative droughts in developed water resource systems.

There exist various tools for the analysis of the management of water resource systems. Some consist of specific models specially developed for the study of a particular system (Shelton, 1979; Palmer et al., 1980; Johnson et al., 1991; Levy and Baecher, 1999; Wagner, 1999; Basson and Van Rooyen, 2001; Cill, 2001; Newlin et al., 2000; Langmantel and Wackerbauer, 2002; Stokelj et al., 2002), and there are also tools designed to be applicable to models of different systems. Among the latter, importance can be given to modules based on the programming of flow networks, which are widely used and accepted because they incorporate optimization techniques in their algorithm systems, among which we could mention the following models: SIMLYD-II, SIM-V, MODSIM, DWRSIM, WEAP, and CALSIM (Everson and Mosly, 1970; Martin, 1983; Labadie, 1992; Chung et al., 1989; Grigg, 1996; DWRC, 2000). Also classified here are the models OPTIGES and SIMGES (Andreu, 1992; Andreu et al., 1992), which are included in the decision support system Aquatool (Andreu et al., 1996) and which were used for the work described in this chapter.

6.5   *Methodology of the analysis*

The experience of IIAMA-UPV during several decades of work on water resource systems analysis has been that integrated management models of water resource systems (WRS) are the best tools to determine the possibilities of experiencing future operative droughts in a WRS and also for determining the effectiveness of the most suitable mitigation measures to be put into practice.

We now examine the details of the methodology used systematically for the analysis of operative droughts and mitigation measures in WRS in the area of the Mediterranean basins in the region of Valencia. These basins are managed basically by two basin agencies: the Hydrographical Confederation of the River Júcar and the Hydrographical Confederation of the River Segura. In order to create the corresponding decision support systems (DSS) the software Aquatool (Andreu et al., 1996) was used, designed by IIAMA-UPV precisely for the development of DSS in the aspect of the integrated analysis of WRS and the prevention and mitigation of operative droughts.

Aquatool permits a model to be made of the integrated management of a WRS composed of multiple supply sources, including surface, underground and nonconventional, multiple commercial water consumers, environmental requirements, multiple transport infrastructures, surface storage, and with extraction from and replenishment of aquifers. Also, with Aquatool, not only quantitative aspects can be studied but also those relating to quality, the environment, and the economy. In the following section we describe and summarize the Aquatool software and the DSS created for the analyses of the basins.

The methodology proposed for the analyses consists of the following stages:

1. Identification of the water resource system.
2. Definition and validation of the model of the complete WRS.
3. Use of the complete model to evaluate the propensity of the WRS to operative droughts on a long-term time scale.
4. Identification and definition of possible measures to reduce the propensity to operative droughts (pro-active measures).
5. Use of the complete model to evaluate the impact of the proactive measures in the indicators of propensity to operative droughts. Following this analysis, those in charge of decision making will select the measures to be applied, taking into consideration, as well as technical criteria (including economic and environmental), the social and economic aspects.
6. Implantation of the measures considered to be the most appropriate.
7. Design of emergency plans against drought. An important aspect is the definition of indicators to identify the risk of suffering an operative drought.
8. Keeping a continual watch on the situation in the system in the course of its management. This must be performed by means of continuous observation of the above-mentioned indicators.
9. Use of the full model to determine the possibility of an operative drought in the WRS in the near future, using the actual conditions as starting point. This analysis improves the quality of the information on the actual situation at the time, since it provides estimations of probability that are not obtainable from the more classical indicators described above.
10. Identification and definition of possible short-term operative drought mitigation measures (reactive measures).
11. Use of the full model to evaluate the impact of the reactive measures on the effects of the prospective drought. Also, after this analysis, those in charge of the decision making will select the measures to be applied, taking into consideration not only the technical criteria (including economic and environmental) but also the social and political.

The analysis and drought measures mentioned in points 3, 4, 5, 6, and 7, corresponding to the management phase defined as planning, are put into effect and must be regularly revised to introduce changes as they occur in the many factors over the years. With regard to this, the Spanish water laws assume a revision of the plans for each basin every five years and the Community Water Board every nine years.

The analysis and the measures described in points 8, 9, 10, and 11 correspond to the management phase defined as exploitation (in real time), and they are processes that, in the semiarid Spanish Mediterranean basins must be continual, theoretically every month, although in some cases a less frequent revision would be admissible, provided that the indicators monitoring the situation in the system (later, we will give some examples) do not make a return to the monthly frequency advisable.

There now follows a detailed description of each of the stages mentioned, together with the observations and recommendations derived from the experience of IIAMA-UPV in applying the methodology in their case studies

6.5.1   Identification of the water resource system

In this phase it is necessary to identify each one of the components of the WRS and to determine its properties, behavior, and relation to the other elements in the system. The main objective of identification is to decide which elements must be included in the WRS management model and the way in which each element is to be modeled. Thus, each of the elements considered to be important is included in the complete WRS management model by means of a "submodel" or "object" related to and interacting with the submodels and objects corresponding to the other elements. In practical terms, the typical elements that comprise a WRS can be grouped as follows:

- Sources or supplies of natural water: This element represents the part of the basin that produces water by natural and renewable means, all of which originally proceed from precipitation and, through hydrological processes, finally appear as some kind of surface water or in the form of a spring.
- Aquifers: Each mass of underground water that forms part of a WRS and that can be managed through pumping or artificial replacement is represented as an aquifer. It is generally difficult to determine the limits of an aquifer, since they are hidden from view, which means that for the purpose of water management estimations they have to be made of their characteristics.
- Natural watercourses: This element represents the natural hydrographical network of a WRS. They have various functions in the management model, the most important of which are to serve as a natural means of movement of water and to represent the necessities of ecological water supplies in rivers.
- Artificial watercourses: Represented by canals, pipes, or other artificial means of water supply, they are normally constructed to supply water for industrial purposes.
- Artificial surface storage elements: These are basically reservoirs or water deposits used to store surplus water for future use.
- Artificial underground water extractors: Represented by wells or similar devices to bring underground water to the surface.
- Artificial replenishment of aquifers: Any artificial process used to increase the volume of aquifers: wells, ponds, etc.

- Management and operational procedures of artificial elements: Represented by any criterion, regulation, or legal norm that controls the normal handling procedures of any of the above-mentioned artificial elements.
- Artificial elements of water production: e.g., desalination plants.
- Artificial elements for the reuse of urban wastewater.

The identification of each one of the above-mentioned elements often requires a careful study in which not only quantitative hydrological aspects must be taken into consideration, but also those relating to quality, society, the economy, and the environment. In this way, the characterization must cover all those aspects relevant to a postdrought analysis, its effects, and the effects of the mitigation measures. From this identification the form of the representation of the element in the model must be decided from a range of possibilities extending from the simple to the complex, establishing a balance between the complexity of the model chosen, the data requirements, a representation sufficiently realistic to provide relevant information on the behavior of the element and its interaction with the rest of the elements in the system. This latter aspect is extremely important. The individual identification of the elements is often difficult precisely because of a high degree of interaction, and a joint identification has to recur in order to achieve some degree of accuracy (see the example of the identification of the surface and underground resources in the Júcar basin and also in that of Turia).

Consequently, during the identification phase, it may become necessary to design specific models to evaluate the behavior of the elements. These specific models are not necessarily the same as those that will later be incorporated in the full model of the WRS, since in many cases complex models are used in the identification phase and simpler ones in the complete model of the system, so that the final models include essential aspects of the more detailed specific models. For example, the specific models developed for the identification phase of the analysis of the water resources in the region of Valencia are described in the following paragraphs.

6.5.1.1 Precipitation-runoff models

The determination of water volumes in natural watercourses at different points of a basin to identify natural water sources is complicated in basins with developed WRS since the artificial actions alter the natural processes and the variations observed at gauging stations, or the water quality may not be representative of the hydrological sector in question. To obtain these variables in their natural state, they have to be recalculated by means of an equation to eliminate the effects of artificial actions. This often implies that it is necessary to know the values of such actions and those of the effects they produce, which is not usually the case. So, the alternative is to use the precipitation-runoff models, which, from the precipitation data, are able to reproduce with more or less detail the stages of the hydrological cycle to obtain the values of water volumes and other variables of interest as they would have been in a completely natural system. In the case of the analyses described in this chapter, SIMPA (Ruiz et al., 1998) was used, to which was added a series of improvements (Pérez, 2004). Therefore, at this moment in time we have available precipitation-runoff models for the following basins or sub-basins: that of the Júcar (Herrero, 2002), Turia (Pérez, 2000), Marina Baja (Gandia, 2001), and Mijares (Sopeña, 2002), whose works are summarized below.

6.5.1.2   Underground flow models

To determine how an underground mass of water functions and its relation with the surface water requires hydrogeological studies in which the geological characteristics of the aquifer are identified, as well as its hydrodynamic qualities, as, for example, hydraulic conductivity, transmissivity, coefficients of storage, the definition of replenishment zones, and other features such as permeability, connections with surface water (rivers, lakes, and reservoirs), and in the case of aquifers near the coast, their connection with the sea. For a correct estimation of the response of the aquifer to various exterior actions (either by human actions or other elements related to the aquifer) that could affect it under normal circumstances or in drought, it may be advisable to construct a distributed model composed of different finites or finite elements. The parameters and conclusions derived from such a model would be useful for the inclusion of the element in the complete management model of the WRS, either by including the aquifer by means of a distributed model or by simpler models that accurately represent the characteristics of the complex model. As is described in the appropriate section, with the Aquatool method it is possible to include aquifers by means of different "submodels" or "objects" of varying complexity according to the data available and the role of the aquifer in the management of the basin and the degree of detail desired in the results. In the cases of the basins analyzed, it was necessary to perform hydrogeological studies and distribution models for the following aquifers: Plana Sur de Valencia, in the basin of the Júcar and aquifers of Sinclinal de Calasparra, Molar, and Vega Alta in the Segura basin. The models were constructed, calibrated, and validated using the software Visual Modflow (Anderman and Hill, 2000). In each of the cases a different solution was reached for its inclusion in the complete basin management model. In the case of the aquifers of Plana Sur and of Molar it was considered sufficient to include them as a unicellular model, while in the case of Sinclinal de Calasparra and Vega Alta they were included as distributed models with the same parameters and discretization as the model of finite differences but using the autovalues methodology designed by IIAMA-UPV for better computational efficiency, which is very helpful if multiple simulations of the WRS management have to be made, as will be seen later.

6.5.1.3   Mixed models

Mixed models are used for the joint identification of surface and underground resources. As has already been mentioned, there are times when attempts to identify separately the surface and underground subsystems can give unsatisfactory results and give rise to errors in the estimation of total water available. This happens, for example, if there is a considerable artificial demand on an aquifer and also when an aquifer has a replenishment component proceeding from returns from irrigation carried out with surface water. An example of the first case was in the identification of the natural sources of supply to a stretch of the river Júcar (from the Alarcón reservoir to the deposits of Molinar), and of the second, on the lower stretch of the Júcar (Alvin, 2001). In both cases it was necessary to resort to mixed models in which the results of the SIMPA precipitation-runoff model were used simultaneously with those of simplified underground flow models.

6.5.1.4 Models of surface water quality

Since one of the effects associated with both natural droughts and operative droughts is low water levels in rivers, and some of the methods adopted serve to reduce water quality, it is important to be able to use tools that allow us to follow the evolution of the quality in basins suffering a drought. In order to identify the aspects of quality in a river it is advisable to create and calibrate specific quality models. In the case of the basins analyzed by IIAMA-UPV, the determination of the evolution in water quality in the lower stretch of the river Júcar was important. Specific models for each of the seven substretches into which the lower reaches of the river were divided were created and calibrated by means of the application of the QUAL-2E (Brown and Barnwell, 1978). The parameters and conclusions obtained (Rodríguez, 2004) were used in the quality model for all the water resources of the Júcar, of which the lower course forms a part.

6.5.2 Definition and validation of the complete model of the system of water resources

This is achieved through the design of a scheme of the system, defining and interconnecting the "objects" or "submodels" chosen to represent each of the a forementioned elements. For this phase the assisted graphic design system of Aquatool was found to be very useful, as it facilitated the insertion of georeferenced factors of the elements in the graph of the scheme, the selection of the model type, access through the graph to the database registers and also their edition, as well as producing written reports on the data entered. It may be said that the graphic interface of Aquatool acts as a specific Geographic Information System for WRS. The elements relative to the definition of the rules of operation are especially important in the design of the model. For this, various mechanisms are available, which may be summed up as: deciding priorities of storage zones in surface reservoirs, priorities in use, priorities of environmental requirements, the definition of alarm mechanisms and the corresponding modifications in supplies, and activation of drought wells. The calibration of priorities and other mechanisms is an important subject. The model is validated by verifying that the resulting management is in accordance with the expected results after the definition of all these management mechanisms.

When the model of the WRS is operative, the behavior of the system in any given scenario can be simulated with any alternatives in the infrastructure, water uses, environmental conditions, and rules of operation.

A hydrological scenario corresponds to a sequence of simultaneous natural inputs at different selected points of a basin for a given time scale. This requisite of naturalization is essential, since otherwise a homogeneous base for the comparison of the effectiveness of measures would not be obtained.

One of the important scenarios, and one which ought always to be borne in mind, is the historic scenario, or historic inflows, corresponding to supplies observed in the system in the past but restored to natural processes as the historic commercial or agricultural activities are gradually abandoned. This historic scenario is normally the one used during the calibration and validation phase of the model.

6.5.3   Use of DSS to determine propensity to operative drought in a water resource system

As has been mentioned, when the operative WRS management model is available, the behavior of the WRS in a future hydrological scenario can be determined. If we were able to predict the hydrological future, and therefore the future water supplies, the analysis would be completely deterministic, and we could simply use the model with known future values, we could estimate the consequences of an operative drought, and then apply steps 4 and 5 (identification measures and evaluation of their efficacy). Unfortunately, the future is usually an unknown quantity in planning (the useful life of infrastructures for established water uses, for example, is around 25 to 50 years).

In the situation of not knowing the hydrological future, various measures can be adopted, the most important of which are the following:

- Use the historic hydrological scenario as the test scenario. In this case, if the series of historical supplies (at different points) are sufficiently long, it can be assumed that something similar will happen in the future in the system, and that the conclusions of the analysis, in terms of the indicators of propensity to drought, are approximations to the real (unknown) values of these indicators, as will be seen later. This option is the most commonly used, in spite of the fact that it is not the best from the statistical point of view to determine the uncertain hydrological future and its consequences. On the other hand, the analysis of the behavior of the WRS, or of any alternative, including the mitigation measures in the following section, in the light of the historic series, is inevitable, since this is an immediate question (What would be the behavior of the system, or of this alternative, if we had a future scenario identical to the historic?). It is advisable to have an answer.

- Use scenarios with possibilities of happening in the future. Since it is improbable that the historic scenario will be repeated in the future, and that the conclusions reached with its simulation are, from a statistical point of view, merely a creation of the population which produced it, it would be good to know the behavior of the system and the mitigation measures, in many other future scenarios, each one with no possibility of becoming reality (as is the case with the historic scenario), but each one with the same probability. With all these combined they give us better approximations to the future drought propensity indicators. All these scenarios proceed from a synthetically generated supply model whose parameters have to be estimated from the statistical properties of the historic series. Aquatool has a module that enables the identification, calibration, and validation of such models from the data of the historic series, as well as the generation of "synthetic" series that can be used as future scenarios. The Mashwin model (Ochoa et al., 2004) creates these "stochastic" models using a traditional approach (ARMA models) and a more novel approach (neuronal networks), the latter developed in IIAMA-UPV (Ochoa-Ribera et al., 2002).

After the historic series, or all the synthetic series, have been simulated the next step is to estimate the operative drought propensity indicators. Since operative droughts happen when any of the users or requirements experiences a deficit, it is possible to obtain custom-made indicators for each one. The most commonly used indicators for the propensity of an element in a system to suffer deficits are (Loucks et al., 1981):

- Guarantee. This is defined as one minus the probability of suffering a deficit, expressed as a percentage.
- Resilience. Defined as the expected duration in time of the deficit.
- Vulnerability. Defined as the total volume of the deficit throughout the drought.

Although these are the theoretical definitions, and, as has been said before, the results of the simulations of a unique series such as the historic, they provide a rough idea of some of these indicators. Aquatool incorporates the calculation of the most widely used indicators.

If the values of the above indicators are such as to warn of a high propensity to operative droughts in all or some of the elements in the system, then this is the moment to think about taking measures to reduce this propensity and to evaluate them through the use of DSS.

In the same way, the DSS tools can be used to evaluate the environmental and economic aspects of the management to achieve a more complete evaluation of the effects of droughts in each of the hydrological scenarios considered. Aquatool also has tools for the analysis of these aspects for an entire basin.

6.5.4 *Identification and definition of possible measures for reducing the propensity to operative droughts (pro-active measures)*

Depending on the WRS and its surroundings and social, economic, environmental, and technical factors, there are many measures available to reduce the propensity to operative droughts. The following are worthy of mention (not necessarily in order of preference):

- Rationalization of the demand: Water uses are often not designed in the most efficient manner possible, so that improvements either in technology or in management can produce savings while they provide the same service
- Direct reutilization of treated effluents
- Improved treatments of effluents
- Increasing the storage capacity of surface water
- Increasing the supply from underground sources
- Desalination plants
- Improvements in the network to reduce losses from pipes, etc. (basin infrastructure)
- Provision of supplies from outside the basin

Together with the above measures, which have a greater or lesser structural factor, it is necessary to consider other measures with less structural impact, but are no less important, such as drawing up a set of rules of operation for the system. The performance of a WRS and the indicators of behavior in a drought depend to a large extent on the operating policies involved in its management, besides the hydrological factors, infrastructure, and the established uses. The optimization of operations in the system must be sought through the drawing up of rules of operation that take into consideration:

- Integrated utilization of all supply sources, and, especially in the Mediterranean basins of Valencia, the combined use of surface and underground water.
- Anticipation of droughts in such a way that the indicators of the hydrological situation allow water-saving measures to be applied in time to avoid extreme emergencies.
- The making of specific rules of operation for each of the pilot systems studied was given special importance. A compilation of the main features of the methodology used can be seen in Solera (2004).
- The establishment of mechanisms for the interchange of supplies among users, so that the water use is optimal from the economic point of view. In this way the economic vulnerability of a system in an operative drought can be greatly reduced. Pulido (2004) contains information on calculating the optimal economic use in a free market,

so that the optimum assignation of supplies can be evaluated and also the desirability of applying management measures in this direction.

- The establishment of other nonstructural measures that could give long-term results, such as citizen education in saving water, changing crops to those that need less water, reducing irrigation by changes in agriculture.

6.5.5 Use of the complete model to evaluate the impact of pro-active measures on the operative drought propensity indicators

The effect of each of the measures mentioned in the foregoing section on the reliability, resilience, and vulnerability indicators of the system in an operative drought are calculated by means of the simulation of the corresponding alternatives using the complete model in the same way as was used in section 6.5.3.

In this way the combination of the most appropriate measures to minimize the propensity of the system to operative droughts can be determined. This combination will have to be a balance between firm antidrought measures and other economic, social, political, and environmental considerations.

In the cases analyzed, different management options were evaluated that had been chosen according to the special needs of each case. Included among these were improvements in the joint use of surface and underground water, the drawing up of rules for the joint operation of reservoirs, and the creation of various measures in anticipation of droughts, which consisted of the programming of precautionary water storage when supplies permitted.

6.5.6 Application of the selected measures

The results obtained from the foregoing measures provide the information necessary for determining the effectiveness and consequences of the possible decisions. Those responsible for the management of the basin will be mindful of these results as well as any other social or political aspects to justify and apply the most appropriate measures.

6.5.7 Design of emergency plans against droughts

One important aspect is the definition of indicators to identify the possibilities of experiencing an operative drought and of the appropriate precautionary measures to reduce its impact. These precautionary measures must be planned in advance, keeping in mind that a balance must be reached between their cost and the real risk of the drought occurring.

In the cases analyzed some drought indicators have been calculated based on the volume of reserves in reservoirs and also on certain precautionary measures consisting of the restriction of the supply of surface water to demands that have at their disposal additional sources of supply such as water from underground.

6.5.8   Permanent monitoring of the situation in the system during its operation

Monitoring must be carried out through continual observation of the indicators in the previous section. For this, basin authorities normally have fairly complicated devices for measuring volumes in rivers and canals, water levels in reservoirs, and rainfall, among others. These data can serve as partial indicators to the situation in the system to a greater or lesser extent.

However, to obtain general information on the state of the system it is necessary to complete the information with a full analysis of the state of the system that correlates all the different factors. In the following section, a method for this type of analysis is proposed.

6.5.9   Use of the complete model to determine the possibility of an operative drought in the WRS in the near future based on the actual situation

This analysis improves the information on the actual present situation since it provides probability estimates unobtainable from the more classical indicators of the previous section. The probability estimates consist of the calculation of the expected value in the coming months of the degree of fulfillment of the forecast supply objectives. The fulfillment of objectives can be evaluated either as supplying the total demand or as different levels of shortfall in the supply.

As has been mentioned previously, Aquatool has a Simrisk module for the simulation of management with multiple synthetic series that provide the statistical results of the simulation. For the evaluation of the short-term operative drought risk this model is used with simulations that begin on the day of the decision making with a duration of one, two, or more years (depending on the "memory" span of the system). The results of the model give an idea of the risk of an operative drought in the ensuing months. If this risk is high, it will be necessary to take measures to mitigate the effects of the possible drought.

6.5.10   Identification and definition of possible measures to mitigate the effects of a possible short-term operative drought (reactive measures)

The measures that can be adopted to mitigate the effects of a possible drought are diverse and also depend on the particular conditions in each basin. They are the measures that, for whatever reason (high cost, infrequent use, etc.), have not been included in the pro-active measures (point 4). Also, it has to be kept in mind that the time available for putting them into practice is limited. Examples of measures of this type would be the restriction of supplies to lower-cost demands, setting up emergency pumping stations, the activation of a water market, interchange of rights, the construction of emergency connections, etc.

6.5.11 *Use of the complete model to evaluate the impact of the reactive measures on possible drought effects*

Any type of measure under consideration will be easy to define beforehand in the complete model in order to evaluate its effect on the system. If there are various alternatives, each one can be evaluated in the model.

Also, as a result of this analysis, those in charge of decision making will select the measures to be applied, considering not only technical factors (including economic and environmental) but also social and political.

One of the main advantages of the proposed analysis is its capacity for dealing with complex systems, giving an overall picture of the situation in the basin as well as of the individual uses, while most of the previously developed indices are applicable only to a demand or to a group of demands. Thus, the proposed method constitutes an authentic early warning system on the arrival of an operative drought.

6.6 *The Aquatool environment for the development of decision support systems*

This system was designed to be an aid to the management and investigation of water resources. It includes an optimization module, a management simulation module, and an underground water preprocessing module. It also has a set of postprocess modules for different types of analysis such as the financial evaluation of management or that of various environmental and water quality parameters. The system is not specifically for a certain type of basin but is designed for general use since it enables different WRS configurations to be represented through graphic design and the graphic introduction of data. Aquatool is at present being used as a support system in several basin management organizations in Spain.

Continuing with the methodology of the analysis described in the previous section, the Aquatool environment provides the following tools: The first point of the methodology analysis deals with the identification of the WRS in order to formulate a model that represents to the highest degree the processes that are to be studied in the real system. The Aquatool system has models to represent a wide variety of types of elements in the real system. The scheme could include any of the following components:

- Nodes with no storage capacity: These permit the user to include river junctions as well as hydrological inflows, derivations, and inputs.
- Nodes with storage capacity: These are for surface reservoirs and supply information on monthly maximum and minimum values for storage and also on evaporation, filtration, size of outlets, etc.

- Channels: It provides five types of channels:

  1. Channels with no loss into or connection with the aquifer.
  2. Channels with filtration losses into an aquifer.
  3. Channels with hydraulic connection to an aquifer. According to the piezometric levels, the aquifer could derive supplies from the river or vice versa.
  4. Channels of hydraulically limited quantity due to the difference between water levels at its extremes.
  5. Channels with hydraulic connection between nodes or vice versa.

- Consumption demands: For example, irrigated zones or municipal and industrial zones. The data consists of the monthly demand. The demand can be supplied from up to five different points on the surface system, with different irrigation efficiency and with surface returns at different points in the system. In this zone it is also possible to pump water from an aquifer with a given maximum pumping capacity. The user can also assign a priority number to the zone. Different zones with the same priority will belong to the same group of users. The model will attempt to share out the water supply within the group according to the needs of each user.

- Hydroelectric plants (nonconsumption demand): They make use of water, but do not consume any significant quantity. They are defined by the maximum flow capacity and by the parameters necessary to calculate the generation of electricity as well as by their objective monthly volumes.

- Aquifers: Underground water can be included using the following types of models, according to the desired degree of detail or to the data available:

  1. Deposit type. The aquifer has no other outlet apart from the water pumped out.
  2. Aquifer with outlet through a spring.
  3. Aquifer with hydraulic connection to surface water, modeled as a unicellular aquifer.
  4. Aquifer with hydraulic connection to surface water, modeled as a multicellular aquifer.
  5. Distributed model of an aquifer using the autovalue method (Andreu and Sahuquillo, 1987). The method gives the same precision in its results as a model in finite differences, but is much more efficient when included in this type of basin management model.

- Other types of element included are return elements, artificial recharge installations, and additional pumping stations.

Also included is the representation of various management norms or criteria, which makes possible the representation of a management approach with the existing norms and also makes possible the analysis and calibration of norms to improve management efficiency. The elements available for this are as follows:

- Objective reservoir curves of volume and zone: Each reservoir will have a curve defined by the user. Minimum (V_{min}) and maximum (V_{max}) monthly volumes will also be given.
- Relations between reservoirs: Different priorities are defined for each reservoir. As is normal in this type of operational rules (Sigvaldason, 1989), all the reservoirs are normally maintained in the same filling zone, provided this is possible, and those with lower priority are diverted first to minor zones rather than those with greater priority.
- Objective minimum volumes for channels: These are usually ecological channels.
- Objective supplies for zones of demand.
- Water destined for turbines in hydroelectric plants.
- Relations between demands, as supply priorities.
- Relations between channels, also given in priorities.
- Relations between elements: Relative priorities can be defined between demands, minimum volumes, and reservoir storage.
- Alarm indicators: These are management criteria whose function is to reduce water consumption when the reserves of the system, or part of it, are below the limits specified by the user.

With all these mechanisms it is possible to represent almost any complex rule of operation for a system, as has been shown by experience.

For the editing and validating phases of the complete WRS model, Aquatool has an assisted graphic design system (Figure 6.1) that facilitates the

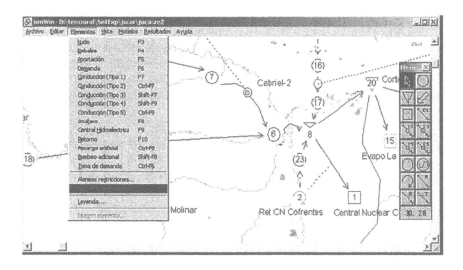

Figure 6.1 Aquatool's graphic interface.

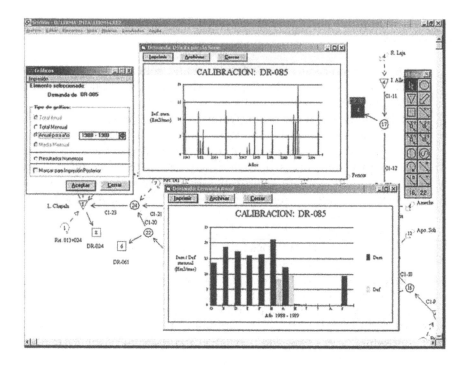

Figure 6.2 Examples of Aquatool's graphic results for a WRS analysis.

georeferenced insertion in the graph of the scheme of each element, selection of the type of model, access through the graph to the database registers and their edition, as well as written reports on the data entered.

For the system's long-term management analysis and the determination of propensity to an operative drought, the Aquatool environment has two optimization and simulation modules (Optiges and Simges). These modules enable the future management of the system for a given hydrological situation to be predicted and provide graphic results of the simulation (Figure 6.2) and the following statistical drought estimators:

- Monthly guarantee: Calculated as one minus the number of months with a failure in the supply divided by the total of the simulated months. It is considered to be a failure when the deficit exceeds a certain threshold.
- Annual guarantee: Similar to the previous case calculated on an annual scale.
- Volumetric guarantee: This is calculated as the quotient between the volume of the supply and the total volume of the demand.
- Maximum monthly deficit: Value of the highest monthly deficit of the simulation.

- Maximum deficit in two consecutive months: As in the previous case calculated for bimonthly periods.
- Utah WRD criterion: The maximum deficit in one year, two consecutive years, and 10 consecutive years is calculated. If any of these three values exceeds a given threshold (different for each period) it is considered that the satisfaction of the demand does not meet this criterion. If all the values are below this threshold, the demand is considered to have been met. If the result of these indicators is "MET" it could be considered that the operative drought risk is acceptable for the element, and if it is "NOT MET" the opposite conclusion is reached.

Aquatool also permits the use of multiple synthetic series, generated by a module for this purpose (Genwin), which uses the stochastic models developed with Mashwin. The Simrisk module is able to simulate all the multiple synthetic series and from the results obtain the following indicators:

- Probability of deficit in any demand element, classified by deficit categories
- Probability of situation of reservoirs, classified in 10 periods per reservoir
- Expected value of the monthly guarantee
- Expected value of the annual guarantee
- Expected value of the volumetric guarantee
- Expected value of the maximum monthly deficit
- Expected value of the maximum deficit in two consecutive months
- Probability of meeting the Utah WRD type criterion in the future simulated time scale

For the evaluation of aspects of water quality in management, Aquatool has a GesCal module (Paredes, 2004). The fundamental characteristic of this tool is the possibility of modeling both reservoirs and stretches of rivers with the same tool and in a way that is integrated with the rest of the elements in the system. Thus, the quality in a stretch of river or in a reservoir does not only depend on the processes involved but also on the system management and on the quality of the different elements related to the element in question. The following constituents can be modeled (Figure 6.3): temperature, arbitrary contaminants, dissolved oxygen together with carbonaceous organic material, nitrogen cycle, and eutrophization. The temperature can even be modeled or included as data of each mass of water.

Aquatool also has tools for the financial analysis of management according to criteria for the financial optimization of management and the evaluation of the cost of water. The following modules are included:

- The Ecoges module for financial optimization of management (Collazos, 2004) evaluates the optimum distribution of water according to

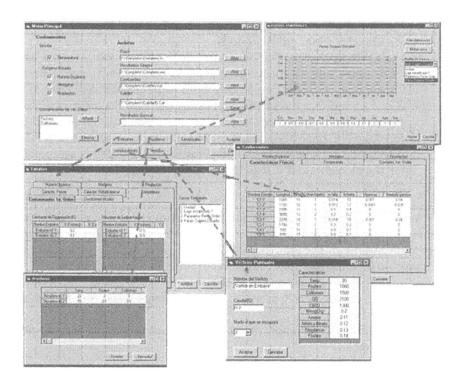

Figure 6.3 GesCal's graphic interface.

market criteria. The program takes into consideration hydrology, infrastructures, or physical conditions and the distribution costs to be paid for the use of water by the users in a certain period of time. Internally, a nonlineal separable function of net profit is optimized for the distribution system.

- The MevalGes module for evaluating water costs at a certain point in the basin, environmental costs, and flood protection costs (Collazos, 2004) gives an estimate of the costs of water as well as the environmental costs by means of the change in value implied in the introduction (or removal) of a resource unit at a certain point at a given time (resource cost), or the change implied in relaxing environmental restrictions in a unit at a certain point and time (or in general at the same time every year). The same concept can be applied to the estimation of other opportunity costs that may be of interest, as, for example, flood protection measures in reservoirs that involve the discharge of water.

The repeated use of the foregoing tools to define the possible measures to be adopted in the real system allows an evaluation to be made of the impact and the effectiveness of each measure. The most appropriate measures proposed for application and emergency plans against drought are thus adequately justified.

For the continuous observation of the situation in the system during operation and the determination of the risk of short-term operative drought, Aquatool has a set of tools that constitutes a complete system of information for the anticipation of droughts. The most important of these tools is the Simrisk module, which permits a simulation to be made of the management with multiple synthetic series and gives the statistical results of the simulation. For short-term operative drought risk evaluation, this module is used with simulations that commence at the time of decision making, with a duration of one, two, or more years (depending on the length of the "memory" of the system). The results provided by the model give an idea of the risk of an operative drought in the forthcoming months. If this risk is high, it will be necessary to take measures to mitigate its effects.

As input scenarios for the multiple simulations any set of scenarios can be used that can feasibly be expected in the near future. Aquatool can utilize the following options:

- The use of one single hydrological scenario taken from the historic series and chosen according to the criterion of the probability of being exceeded. Thus, for example, if a scenario is chosen with a 99% chance of being exceeded, the results can be interpreted as the worst possible future situation.
- Extracting from the historical series the group of scenarios with the same initial month as the scenario under study; thus the expected value for the evolution of the system for a series of feasible future scenarios could be calculated.
- Using the Genwin module for the generation of multiple synthetic series based on the actual situation; in general, the series of natural replenishment of a basin shows a clear time correlation, which is reproduced in the formulation of the classical stochastic models. The use of a synthetic series generation model permits this property of dependence to be utilized by introducing into the model the information on supplies in the previous months so that the generated series is based on the present situation. This makes the series "more probable" than those obtained from the historic series.

For the latter option to be possible, besides obtaining a SAIH (Sistema Automatico de Informacion Hydrologica [automatic system of hydrological information]) that gives the figures of the measured water volumes, a restoration of natural replenishment model has to be formulated, which also automatically gives this data. The Actval model (Andreu et al., 2002) has been developed in Aquatool to make this process automatic. This model was calibrated for the restoration of natural replenishment in the river Júcar.

When the multiple simulations with historic or synthetic scenarios have been completed, the probabilities of a shortfall in the system in the forthcoming months are estimated. The Simrisk module works out the following indicators for short-term management:

- Probability of a monthly shortfall in the supply to a demand: Calculated for each month of the simulation period as the probability of suffering a deficit.
- Probability of monthly shortfall by level of supply: Considering the volume of demand divided into levels.
- Probability of excess in the volume of the shortfall.
- Probability of the monthly situation of reservoirs in levels: For each month of the simulation period, the probability of the reservoir finishing up with a volume of reserves in a given interval.
- Probability of no excess in monthly storage of reservoir.
- Probability of monthly shortfall in the supply of the minimum volume: This is calculated for each month and for each river course.

The previous values provide an estimation of the risk of operative drought in the forthcoming months. If this risk is high, it will be necessary to take measures to mitigate possible effects.

Aquatool also has a set of graphic analysis tools for the results of the foregoing statistics, which provide a thorough evaluation of the figures. They are as follows (Figure 6.4):

- Graphs of the risk of shortfall in the demand: They contain the graphic representation of the risk of the monthly shortfall by supply levels calculated for each demand. The months of the study are shown in the ordinates axis and the percentage probability in the abscissas. The value of the risk of the shortfall happening at each level of demand defined is shown in the form of vertical bars. The highest value of each vertical bar represents the accumulated risk of a deficit occurring of a magnitude greater than the lowest limit of the corresponding interval.
- Graphs of no excess in the deficit: They contain the graphic representation of the statistics of the probability of no excess in the intensity of the deficit. The ordinates axis shows the months of the study and the abscissas the value of the deficit in a monthly percentage. Each curve represents the value of the deficit as a percentage with a given probability of no excess.
- Probability graphs for the state of reservoirs: These contain the graphic representation of the statistics of probability of the monthly state of the reservoir. The months of the studio are on the ordinates axis and the percentage probability on the abscissas. The value of the probability of the reservoir ending the month within each interval of defined volume is given in the form of vertical bars. The highest value of each vertical bar represents the probability of the reservoir finishing the month below the highest value of the interval. When the reservoir

reaches the end of the month completely full, the result is not included in any of the intervals, thus obtaining also a measure of the probability of overflows, which would be equal to 100% of the complement of the sum of the probabilities calculated for all the stretches.

- Probability graphs for no excess in storage: These contain the graphic representation of the statistics of the probability of no excess in the monthly storage of the reservoir. The months of the study are shown on the ordinates axis and the percentage volume of the reservoir in the abscissas. Each curve represents the reservoir value whose probability of no excess is a given value.
- Graphs of probability of excess in a month: These contain the results of the previous graph fitted to a specific month, with the probability of no excess on the ordinates axis and the reservoir volume corresponding to this probability on the abscissas.

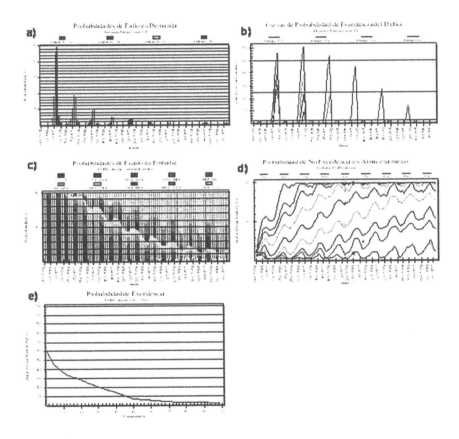

Figure 6.4 Simrisk graphic results: (a) Graphic of monthly risk failure in demands. (b) Graphic of probability to not exceeding the deficit. (c) Graphic of reservoir's level probability. (d) Graphic of not exceeding probability in storage. (e) Graphic of exceeding probability in a month.

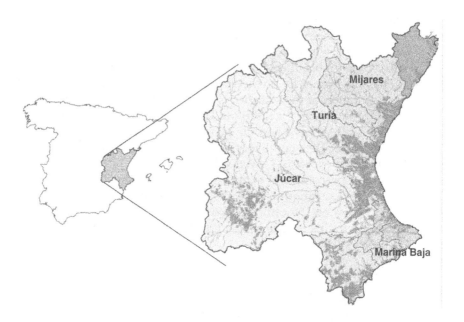

Figure 6.5 Situation map of the studied basins.

The Aquatool user interface also provides different facilities to include in the simulation short-term drought mitigation measures. This means that a rehearsal of the measures can be made in advance and an evaluation made of their efficacy to help in choosing the most efficient measures.

6.7 Case studies

The process described in the methodology of analysis was carried out systematically for the pilot cases using the Aquatool environment. The corresponding detailed reports for each basin were produced as a result of the studies: Júcar, Turia, Mijares, and Marina Baja (Figure 6.5), all of which belong to the management area of the Hydrographical Confederation of the Júcar. In this section the principal characteristics and conclusions obtained from the studies are presented.

6.7.1 System of the Júcar

The area of the Júcar basin lies in the autonomous regions of Castilla la Mancha and Valencia. It occupies an area of around 22,000 km², and the biggest river is almost 500 km long. The average total rainfall in the basin is 510 mm per year. The average total input volume is 1450 hm³ per year, of which renewable underground resources make up an estimated 80%. Supplies to towns and cities reach approximately 150 hm³ per year to a total of

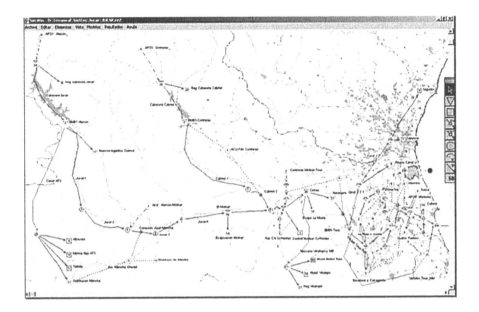

Figure 6.6 Júcar management analysis model.

about 860,000 inhabitants. The area under irrigation is 158,500 h and uses around 1000 hm³ per year. Finally, the established hydroelectric potential is of the order of 1300 MW, although 40% of this (540 MW) is supplied from the reversible station of Cortes-La Muela. The existing reservoirs have a total capacity of approximately 2900 hm³, the most important being the reservoirs of Alarcón and Contreras, situated at the headwaters of the Júcar and the Cabriel and the Tous in the lower basin. The Tous reservoir is situated directly upstream of the areas of greatest demand. Underground resources of the order of 300 hm³ per year are also utilized.

A scheme of the model constructed for the management analysis of the Júcar system is shown in Figure 6.6. The most important zones in this scheme are:

- The La Mancha aquifer: Situated in midbasin, in the past 20 years there has been a drastic increase in its exploitation, which has drastically altered its relation with the river in the form of inflow/outflow. For the construction of this model it was necessary to make on-the-spot specific detailed evaluations of the resources and of their interrelations with the surface system.
- Irrigation zones downstream of the Tous reservoir: They comprise more than 50% of the demand for water in the basin. They also give considerable return surface flows to the river and filtration to the aquifer connected to it as well as to other hydrogeological units, which greatly complicates the construction of a model of this zone.

In relation to the identification of drought situations, the transfer of resources to other basins suffering shortages, such as Vinalopó and Marina Baja, plays an important role.

The management analysis of the system clearly shows the great variety of the system's resources, which can provide both periods of abundant water and others of extreme drought with serious water shortages. This situation underlines the importance of correct management planning to propose and validate effective measures to reduce the vulnerability of the system to droughts, while allowing better exploitation of the resources at other times.

Among these measures, it would be important to drill drought wells in the irrigation zone downstream of the Tous and to lay down the rules of operation for reservoirs proposed in the so-called Alarcón Agreement (MMA, 2001) in which measures were established to save water by reducing the transfers to other basins with shortages.

Use of the Simrisk model for the continuous monitoring of the management of the system was also adopted. Every month, an evaluation was made of the probability of running out of water during the current irrigation campaign, analyzing preventive measures when the risk was considered to be high (Figure 6.4e).

6.7.2   The system of the Turia

The Turia occupies an area of approximately 6400 km² in the autonomous regions of Castilla La Mancha and Valencia. Its southern limits are the river basins of the Júcar and the Poyo and in the north the basins of Mijares, Palancia, and the Carraixet. The average total rainfall in the system is 515 mm annually, with an average temperature of around 14°C. The population of the area is 1,443,914 according to the census of 1991, with a supply of 32 hm³/year. The demand for irrigation water is of the order of 295 hm³/year, of which 85 hm³/year (Camp de Turia) come from both surface and underground water. The reservoir capacity is 328 hm³ in Buseo, Arquillo de San Blas, Benageber, and Loriguilla, all of which are on the principal watercourse.

The scheme of the management analysis of the Turia system is shown in Figure 6.7. In this system the main reservoir is Benageber, with a capacity of 228 hm³, comprises 70% of the supplies from the river and is the principal water resource of the system. The demand of Camp de Turia is also important. This was originally developed from underground resources theoretically unconnected with the river and subsequently improved by including surface water. This solution is especially sensitive, since the surface resources of the river are not of sufficient capacity to guarantee the increase, so that good planning is crucial in order to guarantee supplies to the rest of the system while increasing the surface supply to Camp de Turia.

The analysis of the system shows its high degree of reliability, excluding the surface supply to the demand of Camp de Turia, together with the occasional generation of excess volume in the river. It also shows how the

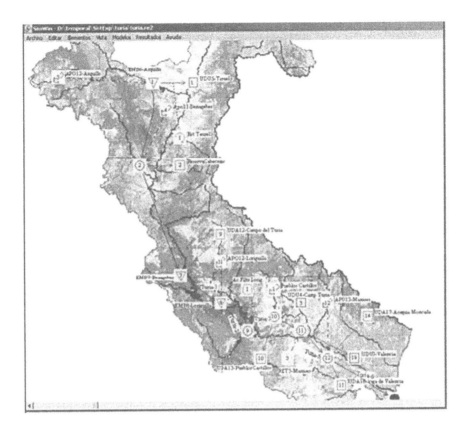

Figure 6.7 Turia management analysis model.

guarantee of supplies is notably reduced when the volume of water diverted to Camp de Turia is increased.

As a corrective measure, a management policy was suggested and analyzed, consisting of impeding the supply of surface water to this demand when the Benageber reserves fall below a certain limit. With this procedure a threshold value was obtained that guarantees supplies to the preferential demands in the system, while maintaining the supply of surface water to Camp de Turia at a high level.

6.7.3 The system of the Mijares

The Mijares basin is shared between the provinces of Teruel and Castellón. It occupies a total surface area of 5466 km² and is comprised of two geographic zones with two distinct climates: one bordering on the sea with a Mediterranean coastal climate, and the other upstream of the Arenós reservoir with a somewhat more continental climate. The average annual rainfall in the zone is 505 mm, and the average temperature 14.4°C. It has a total population of 363,578 inhabitants, according to a 1991 census. The towns

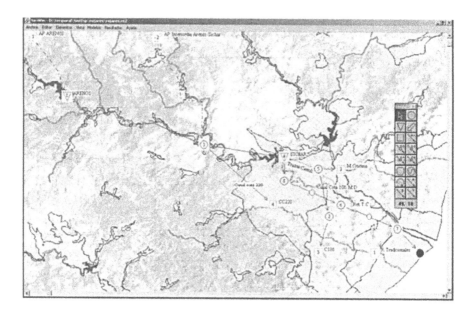

Figure 6.8 Mijares management analysis model.

with more than 15,000 inhabitants are supplied exclusively from wells. The total surface area under cultivation is 124,310 h, of which 43,530 (35%) is irrigated, while the rest (65%) is devoted to nonirrigated crops. Citrus fruit is the chief product, and it occupies about 87% of the irrigated land. There are two reservoirs, Arenós and Sichar, with capacities of 130 and 52 Hm^3 respectively. There are two different administrative areas known as *traditional irrigation* and *mixed irrigation*. The former have preferential rights in the use of surface river water, and the latter are supplied mostly from underground water but are allowed the use of surface water when this can be done without prejudice to the traditional irrigation.

The scheme of the Mijares management analysis is given in Figure 6.8. The most important feature of this system is the great seasonal variation in the river level, which gives rise to alternate periods of serious drought and others of flooding with the reservoirs overflowing their banks. These conditions mean that great thought must be given to the planning of the joint use of surface and underground water, to avoid the overexploitation of aquifers and to make the most of the surface supply. The management analysis includes the rules of operation for the surface supply to the mixed irrigation farms by means of a reserves curve defined by the users' agreement (CHJ, 1973).

The analysis of the system management by means of calculation models permitted, in the first place, the determination of the effectiveness of the present reserves curve, and second, showed the benefits that may be gained through the use of management criteria based on risk estimation

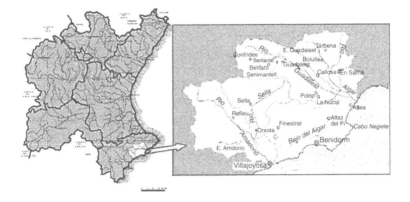

Figure 6.9 Situation map of the basins in the Marina Baja system.

(as suggested in points 9 and successive of the methodology proposed in this chapter).

6.7.4 Marina Baja system

The Marina Baja system is situated in the province of Alicante and consists of the basins belonging to the rivers Algar and Amadorio and the coastal subbasins between the river Algar and the southern limit of the municipality of Villajoyosa (Figure 6.9). It has a total surface area of 583 km².

The climate in the system is semiarid Mediterranean. Average annual rainfall is 400 mm per year, and the average temperature is 16°C. Total population is 137,843 inhabitants, according to the 1991 census, mostly concentrated in coastal areas. In summer, due to the influx of tourists, the population increases by about 225%. The total area under cultivation is 13,581 h, of which 8023 (59%) are irrigated and the rest (41%) are devoted to nonirrigated production. Since the 1979–1985 drought, joint use has been made of surface and underground water and the reutilization for irrigation of recycled urban wastewater. The system is at present in a situation of deficit as regards natural renewable resources in the basin, and attempts are being made to solve the problem by bringing water from the Júcar basin.

The Marina Baja analysis model is given in Figure 6.10. Its most important feature is the exploitation of the Berniá-Ferrer aquifer subsystem through the wells of Algar, which provide more than 50% of the system's resources. This analysis focused on a study of the aquifer with a view to drawing up a set of operating rules for the wells, which would allow the summer demand to be maintained by the timely use of the wells in periods when surface water is scarce.

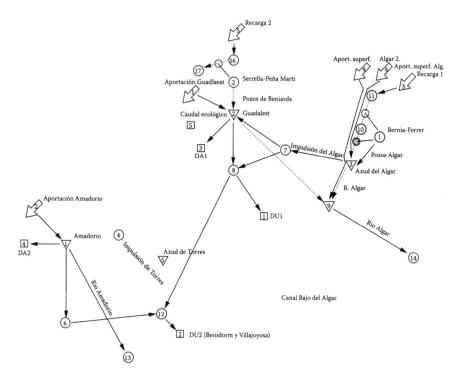

Figure 6.10 Marina Baja management analysis model.

6.8 Conclusion

In this chapter, a generic methodology for the analysis of water resource systems is proposed, whose aim is the design and planning of operational measures that would avoid or mitigate the effects of droughts. A WRS is considered to be a set of interconnected natural and artificial elements in one or more hydrographical basins.

In developed water resource systems, once the water requirements have been determined for different uses, including the environmental, if the water produced by the hydrology and management of the system is insufficient for its needs, it can be said that a condition of operative drought exists, to distinguish the situation from droughts that result from natural conditions only, and also to stress the importance of the correct management of the system in such conditions.

It was shown that the appropriate time scale for the analysis is one month, unless the circumstances required a different period of time, and that the best unit of area was to consider the hydrological basin as a whole.

Throughout various decades of dedication to the analysis of WRS, the experience of IIAMA-UPV has shown that integrated management models of water resource systems are the best tools to determine the probability of suffering a future operative drought in the system, as well as for determining the efficacy of the most appropriate mitigation measures to be put into practice.

Details are given of the methodology used systematically for the analysis of operative droughts and mitigation measures in the WRS in the Mediterranean basins of the region of Valencia. For the creation of the corresponding decision support systems, the software Aquatool was used, which was designed by IIAMA-UPV specifically for developing DSS for the integrated analysis of water resource systems and for the prevention and mitigation of operative droughts. Aquatool allows a model to be made of the integrated management of a WRS composed of multiple sources, both surface, subterranean, and nonconventional, multiple commercial water uses, environmental requirements, and multiple infrastructures for conduits, surface storage, supply to and replenishment of aquifers. With Aquatool, not only the quantitative aspects can be studied, but also the aspects of quality, the environment, and economics.

In conclusion, we can say that the systematic application of this methodology to the basins of Valencia, through the use of decision support systems, has made possible:

- The determination of the risk of operative droughts in the basins, as well as the analysis of the efficacy of different pro-active measures designed to mitigate their effects
- The creation of rules of operation against possible droughts and the adoption of mitigation measures in real time
- Estimation of the probability of suffering an operative drought in the short term, and analysis of the efficacy of the reactive measures to reduce any losses occasioned by it.

6.9 Acknowledgments

We thank the Commission of the European Communities for their financing in the project "Water Resources System Planning, WARSYP," contract ENV4-CT97-0454 (Directorate General XII Science, Research and Development); the project "Water Resources Management Under Drought Conditions, WAM-ME," contract ICA3.1999.00014 (Directorate General XII Science, Research and Development); and the project "SEDEMED — Sécheresse et Desertification dans les bassins méditerranées," 2002-02-4.4-1084 (INTERREG III B-Mediterranée Occidentale).

We also thank the Ministerio de Educación y Cultura de España (Comisión Interministerial de Ciencia y Tecnología, CICYT) for their financing of the project "Desarrollo de Elementos de un Sistema Soporte de Decisión para la Gestión de Recursos Hídricos," HID1999-0656 and the project "Sistema de apoyo a la decisión para la gestión cuantitativa, cualitativa y ambiental de cuencas hidrográficas," REN2002/03192.

Thanks are extended to the Oficinas de Planificación Hidrológica and Áreas de Explotación in the Confederaciones Hidrográficas del Júcar, del Tajo y del Segura and the Centro de Estudios Hidrográficos del CEDEX, for supply the required data for the research and development of the decision support system Aquatool.

We also thank the Foreign Language Co-ordination Office at the Polytechnic University of Valencia for their help in translating this chapter.

References

Albín, J. V. (2001). *Estudio para la determinación de las aportaciones intermedias entre la presa de Tous y la estación de aforos de Huerto Mulet del río Júcar.* Ejercicio final de carrera para la titulación de Ingeniero de Caminos, C. y P. en la Universidad Politécnica de Valencia.

Anderman, E. R., and Hill, M. C. (2000). *MODFLOW-2000, the U.S. Geological Survey modular ground-waste model-documentation of the hydrogeologic-unit flow (huf) package.* Denver: USGS.

Andreu, J. (1992). *Modelo optiges de optimización de la gestión de esquemas de recursos hídricos. Manual de usuario.* SPUPV-92.2012. Valencia, Spain.

Andreu, J. (1993). Análisis de sistemas y modelación. In J. Andreu (Ed.), *Concéptos y métodos para la planificación hidrológica* (pp. 25–33). Barcelona: CIMNE.

Andreu, J., Capilla, J., and Ferrer, J. (1992). *Modelo Simges de simulación de la gestión de recursos hídricos, incluyendo utilización conjunta. Manual del usuario.* SP-UPV-92.1097. Valencia, Spain.

Andreu, J., Capilla, J., and Sanchis, E. (1996). Aquatool: A generalized decision support-system for water-resources planning and operational management. *J. Hydrology* 177, 269–291.

Basson, M. S., and Van Rooyen, J. A. (2001). Practical application of probabilistical approaches to the management of water resource systems. *J. Hydrology* 241, 53–61.

Brown, L. C., and Barnwell, T. O. (1978). *The enhanced stream water quality models QUAL2E and QUAL2E-UNCAS.* EPA/600/3-87/007. Athens, GA: U.S. Environmental Protection Agency.

Buras, N. (2000, March). Building new water resources projects or managing exiting systems? *Water Int.* 25(1), 110–114.

CHJ. (1973). *Convenio de Bases para la Ordenación de las Aguas del río Mijares. Reglamento del sindicato central de regantes del río Mijares.* Editado por la Confederación Hidrográfica del Júcar, Ministerio de Obras Públicas, Transportes y medio Ambiente.

Chung et al. (1989).

CiII. (2001). *Análisis de la gestión de aducción: mes de octubre de 2001.* Edición periódica de la comisión de explotación de captaciones del Canal de Isabel II.

Collazos, G. (2004). *Metodología y Herramientas para análisis económicos de SRH requeridos por la DMA.* Informe parcial de tesis doctoral dirigida por D. Joaquín Andreu en la Escuela Técnica Superior de Ingenieros de Caminos C. y P. de Valencia

DWRC. (2000). *CALSIM water resources simulation model.* Bay Delta, CA: Department of Water Resources. (Also http://modeling.water.ca.gov/hydro/model/index.html).

Everson, D. E., and Mosly, J. C. (1970). Simulation/optimization techniques for multi-basin water resource planning. *Revista Water Resour.* 6(5), 725–723.

Gandia, R. (2001). *Análisis de la situación actual de explotación del sistema de recursos hídricos de La Marina Baja.* Proyecto final de carrera para la titulación de ICCP presentado en la ETSICCP de la UPV. Dirigido por A. Solera.

Grigg, N. S. (1996). *Water resources management: Principles, regulations, and cases.* New York: McGraw-Hill.

Herrero, R. (2002). *Gestión del sistema de la cuenca del Júcar basada en riesgo de sequías, con revisión de aportaciones aguas arriba del embalse de Tous.* Ejercicio final de carrera para la titulación de ICCP en la UPV, Valencia.

Ito, K., Xu, Z. X., Jinno, Z., Kojiri, T., and Kawamura, A. (2001, July–August). Decision support system or surface water planning in river basins. *J. Water Resour. Planning Man.* 127(4), 272–276.

Johnson, S. A., Stedinger, J. R., and Staschus, K. (1991, May). Heuristic operating policies for reservoir system simulation. *Revista Water Resour. Res.* 27(5), 673–685.

Labadie, J. W. (1992). *Generalized river basin network flow model: Program MODSIM.* Fort Collins, CO: Department of Civil Engineering, Colorado State University.

Langmantel, E., and Wackerbauer, J. (2002, November). *A regional model of economic development and industrial water use in the catchment area of the upper Danube.* Conferencia Internacional De Organismos De Cuenca, Madrid.

Levy, B. S., and Baecher, G. B. (1999, March–April). NileSim: A Windows-based hydrologic simulator of the Nile river basin. *J. Water Resour. Planning Man.* 125(2).

Loucks, D. P. (2000, March). Sustainable water resources management. *Water Int.* 25(1), 3–10.

Loucks, D. P., Stedinger, J. R., and Haith, D. A. (1981). *Water resource systems planning and analysis.* Englewood Cliffs, NJ: Prentice Hall.

Martin, Q. W. (1983). Optimal operation of multiple reservoir systems. *J. Water Resour. Planning Manage.* 106(1), 58–74

MMA. (2001, July). *Convenio específico sobre el embalse de Alarcón para la gestión optimizada y unitaria del sistema hidráulico Júcar (Alarcón Contreras y Tous).* Celebrado entre e Ministerio de Medio Ambiente y la Unión Sindical de Usuarios del Júcar USUJ.

Newlin, B., Jenkins, M. W., Lund, J. R., and Howitt, R. E. (2000). Southern California water markets: Potential and limitations. *J. Water Resour. Planning Manage.*

Ochoa-Ribera et al. (2002). Multivariate synthetic streamflow generation using a hybrid model based on artificial neural networks. *HESS — Hydrological and Hearth Sys. Sci.* EGS 6(4), 641–654.

Palmer, W. C. (1965). Meteorologic drought. *Res. Pap. U.S. Weather Bur.* 45(58), 19–65.

Palmer, N. R., Wright, J. R., Smith, J. A., Cohon, J. L., and Revelle, C. S. (1980). *Policy analysis of reservoir operation in the Potomac river basin:* Vol. I, Executive summary. Baltimore: Johns Hopkins University Press.

Paredes, J. (2004, December). *Integracion de la modelación de la calidad del agua en un sistema de ayuda a la decisión para la gestión de recursos hídricos.* Tesis doct. dirigida por D. Joaquín Andreu para la tit. de Dr. ICCP.

Pérez, M. A. (2000). *Modelación cuasidistribuida de los recursos hídricos y estudio de explotación del río Túria.* Proyecto final de carrera para la titulación de ICCP presentado en la ETSICCP de valencia. Dirigido por Abel Solera.

Pérez, M. A. (2004). *Análisis de presiones e impactos mediante sistemas de información geográfica en cuencas hidrográficas. Aportación a la directiva marco de las aguas.* Tesis doctoral. Universidad Politécniccca de Valencia, Spain.

Pulido, M. (2004, February). *Optimización económica de la gestión del uso conjunto de aguas superficiales y subterráneas en un sistema de recursos hídricos. Contribución al análisis económico propuesto en la Directiva Marco Europea del agua.* Tesis doct. dirigida por D. Joaquín Andreu para la tit. de Dr. ICCP.

Rodriguez, A. (2004). *Estudio de mejora de la calidad del agua en la cuenca del río Júcar aguas abajo del embalse de Tous.* Ejercicio final de carrera para la titulación de Ingeniero de Caminos, C. y P. en la Universidad Politécnica de Valencia. Dirigido por: J. Paredes y J.Andreu. Valencia.

Ruiz, J. M. (1998). *Desarrollo de un modelo hidrológico conceptual-distribuído de siulación contínua integrado con un Sistema de Información Geográfica.* Tesis doctoral dirigida por D. J. Andreu y D. T. Estrela. Presentada en la ETSICCP de la UPV para la titulación de Dr. Ing. De CCP.

Sánchez, S. T., Andreu, J., and Solera, A. (2001). *Gestión de Recursos Hídricos con Decisines Basadas en Estimación del Riesgo.* Valencia: Universidad Politécnica de Valencia.

Shelton, A. R. (1979). *Management of TVA reservoir systems.* Proceedings of National Workshop on Reservoir Systems Operations. Boulder: University of Colorado.

Sigvaldason, O. T. (1989, September 18–29). *Simulation models for representing long-term and short-term system operation.* Proceedings of NATO Advanced Study Institute, Stochastic Hydrology in Water Resources Systems: Simulation and Optimization, Peñíscola, Spain.

Solera, A. (2004). *Herramientas y métodos para la ayuda a la decisión en la gestión sistemática de recursos hídricos. Aplicación a las cuencas de los ríos Tajo y Júcar.* Tesis doctoral dirigida por J.Andreu. Presentada en la ETSICCP de la UPV para la titulación de Dr. Ing. de CCP.

Sopeña, F. J. (2002). *Análisis del sistema del río Mijares y diseño de un plan de gestión óptimo para la mitigación de sequías.* Proyecto final de carrera para la titulación de ICCP presentado en la ETSICCP de valencia. Dirigido por A. Solera

Stokelj, T., Paravan, D., and Golob, R. (2002, November). Algorithm for run-of-river hydropower plants. *J. Water Resour. Planning Manage.* 128(6),   .

U.S. Army Corps of Engineers (1966, 1975). *Program description and user manual for SSARR model, streamflow synthesis and reservoir regulation.* Portland, OR: North Pacific Division.

Vlachos, E., and James, L. D. (1983). Drought impacts. In V. Yevjevich et al. (Eds.), *Coping with droughts* (pp. 44–73). Water Resources Publications.

Wagner, A. I. (1999, October 27–29). *Aspectos sobre la política de operación de presas en los reglamentos de los sistemas de riego.* Presentado el IX congreso nacional de irrigación: Simposio 6 Reglamentación de Sistemas de riego. Culiacán, Sinaloa, México.

Wilhite, D. A., and Glantz, M. H. (1985). Understanding the drought phenomenon: The role of definitions. *Water Int.* 10, 110–120.

chapter seven

Droughts and the European water framework directive: Implications on Spanish river basin districts

Teodoro Estrela
Confederación Hidrográfica del Júcar, Spain

Aránzazu Fidalgo
Confederación Hidrográfica del Júcar, Spain

Miguel Angel Pérez
Universidad Politécnica de Valencia, Spain

Contents

7.1   Introduction

The European Water Framework Directive (WFD) (2000/60/EC) establishes a framework for community action in the field of water policy. The main objective of the WFD is to achieve the good status of water bodies, protecting them and impeding their deterioration. This directive represents a substantial change in the traditional approach for water management since:

- It emphasizes water quality aspects, environmental functions, and a sustainable water use, contributing to mitigate the effects of floods and droughts.
- It establishes the river basin as the basic unit for water management including in its domain groundwater, transitional, and coastal waters.
- It requires transparency in the access to hydrological and environmental data, forcing standardization of procedures to determine the environmental status of water bodies.
- It introduces the principle of cost recovery favoring a greater public participation in the whole process.

The WFD is a complex directive that imposes a large number of tasks on European Union member states. The directive is organized into 53 statements, 26 articles, and 11 annexes, which is transferred to the legal system of member states.

A key aspect of the WFD implementation has been the creation of a network of European pilot river basins with the main goal to ensure the coherence and crossed application of the guide documents elaborated by working groups made by experts from the member states. Spain assumed the highest level of compromise by proposing verification and evaluation, in the territorial area of the Júcar River Basin Authority (RBA), which is one of the pilot river basins, of all guide documents and agreed to work on the development of a platform of a common Geographic Information System.

In this chapter droughts are analyzed from the perspective of the WFD, placing emphasis on drought planning and management aspects and focusing on the case of Spain and more specifically on the Júcar RBA.

7.2   Droughts in the WFD

Droughts are considered in different statements, articles, and annexes of the WFD. Statement 32 states:

> There may be grounds for exemptions from the requirement to prevent further deterioration or to achieve good status under specific conditions, if the failure is the result of unforeseen or exceptional circumstances, in particular floods and droughts... provided that all practicable steps are taken to mitigate the adverse impact on the status of the body of water.

In Article 1 (Purpose), the purpose of the directive is specified to establish a framework for the protection of inland surface waters, transitional waters, coastal waters, and ground water, which prevents their further deterioration, protects and enhances the status of aquatic ecosystems, promotes sustainable water use, aims at enhance protection and improvement of the aquatic environment by promoting a progressive reduction of discharges, ensures a continuing reduction of pollution of ground water, prevents its further pollution, and contributes to mitigate the effects of floods and droughts.

Point 6 of Article 4 (Environmental objectives) explains that temporary deterioration of the status of water bodies shall not be in breach of the requirements of this directive if this is the result of circumstances of natural cause or force majeure, in particular extreme floods and prolonged droughts, when all of the following conditions have been met : (a) all practicable steps are taken to prevent further deterioration in status, (b) the conditions under which circumstances that are exceptional or that could not reasonably have been foreseen may be declared, including the adoption of the appropriate indicators, are stated in the River Basin Management Plan, (c) the measures to be taken under such exceptional circumstances are included in the program of measures, and (d) a summary of the effects of the circumstances and of such measures taken or to be taken is included in the next update of the River Basin Management Plan.

In Annex 6 (Lists of measures to be included within the programmes of measures) Part B the demand management measures are included, which describe inter alia the promotion of adapted agricultural production, such as low water requiring crops in areas affected by droughts.

To summarize:

- Droughts constitute an exemption from some WFD requirements.
- The declaration of a drought situation must be defined in the Basin Management Plan, adopting adequate indicators.
- Measures to be adopted in drought situations must be incorporated in the Programme of Measures.
- The Basin Management Plan, once updated, will summarize the effects of droughts and measures.
- Low water requiring crops should be applied in areas affected by droughts.

7.3 Drought planning legal framework in Spain

Drought management can be carried out by two main approaches:

1. As an emergency situation, that is considering it as a crisis situation, which can be restored with extraordinary water resources.
2. As a current element of the general water planning and management, which means that a risk analysis must be carried out to assess its probability of occurrence and measures to be applied must be planned ahead.

In Spain, droughts have been traditionally managed according to the first approach, although since the entry into force of the Hydrologic National Planning Act (HNP, 2001) both approaches should be used.

The Water Act foresees proper measures for strong drought situations. These measures are determined by the Spanish government and are focused on the use of the public hydraulic domain. They are submitted by the so-called Royal Decree Acts of urgent exceptional measures. Public works (mainly drought wells) that result from these measures are declared of public use and the private property where they might be located can be expropriated for immediate construction.

Clear examples are the urgent measures applied at the beginning of the 1980s or during the years 1994 and 1995, with the building of urban supply pipes. Examples of laws associated with urgent measures for drought situations are:

- Act: "Ley 6/1983 de 29 de junio de 1983, sobre medidas excepcionales para el aprovechamiento de los recursos hidráulicos escasos a consecuencia de la prolongada sequía"
- Act: "Ley 15/1984 de 24 de mayo, para el aprovechamiento de los recursos hidráulicos escasos a consecuencia de la prolongada sequía"
- Act: "Real Decreto-Ley 8/2000, de 4 de agosto, de adopción de medidas de carácter urgente para paliar los efectos producidos por la sequía y otras adversidades climáticas"

The formal procedures of response to droughts should be considered in a more integrated planning for the coming years. Article 27 of Act 10/2001, July 5, of the National Hydrologic Plan (NHP) refers to drought planning, stating in point 1:

> For the intercommunity basins, the Ministry of Environment, in order to minimise the environmental, economic, and social impact of any situations of drought, shall establish an overall system of water indicators that allows these situations to be predicted and acts as a general reference for Basin Organisations to formally declare situations of alert and temporary drought. This declaration shall involve the implementation of the Special Plan described in the following point.

Also point 2 of the same article specifies:

> Basin Organisations shall draw up, in the scope of the corresponding Basin Hydrological Plans, and within the period of two years from this Act coming into force, special action plans in situations of alert and temporary drought, including the rules for exploitation of systems and the measures to implement with relation to the use of the public water domain. The mentioned plans, subsequent

to a report from the Water Council for each basin, shall be sent to the Ministry of Environment for their approval.

Finally, point 3 of the referred article 27 states:

> The Public Administrations responsible for urban supply systems, which serve, singly or jointly, towns of 20,000 inhabitants or more, must have an Emergency Plan for drought situations. This Plan, which shall be reported by the Basin Organisation or corresponding Water Authorities, must take into consideration the rules and measures laid down by Special Plan mentioned in point 2, and must be operative within a maximum period of four years.

7.4   Drought management tools

Drought situations are extreme hydrological events where water is scarce, and precipitation is at a minimal level. They are characterized by having long duration with starting and ending periods uncertain.

The anticipation in the application of mitigation measures becomes an essential tool for the reduction of socioeconomic effects of droughts; that is why having completed indicators systems that allow early warning of these extreme events is essential. These systems must be considered as key elements in drought events management and in the strategic planning of the actions to be taken.

The main tools for drought management and planning available in Spain are:

- Drought indicators for the Spanish territory
- Drought indicators for the River basin district
- The River Basin Drought Special Plan
- The Emergency Plan for public water supplies greater than 20,000 inhabitants

These tools are described in the following section.

7.5   Drought indicators for the Spanish territory

Currently, a Spanish Indicator System has been established in order to assess the quantitative status of water resources in the different exploitation systems existing in each river basin district. The Spanish Ministry of Environment has done this task jointly with the Centre of Studies and Experimentation of Public Works (CEDEX).

Different parameters have been chosen (inflows, outflows, and storage in reservoirs, flow river gauges, precipitation, and aquifer water level) for each exploitation system. These parameters are used to assess the quantitative status of water resources in each system, comparing the record achieved

Sistema de Indicadores Hidrológicos
TIPO
• Precipitación
• Caudales aforados
▲ Entradas en Embalses
* Resenas de Embalses
◦ Niveles Piezométricos
★ Salidas en Embalses

Figure 7.1 Tentative points of the Spanish Drought Indicator System.

in a determined period of time that has a historical and representative mean value. Figure 7.1 shows the location of the selected control points.

The comparison is expressed in terms of different percentages depending on the adopted temporal period of analysis (one month, three accumulated months, or 12 accumulated months). Figure 7.2 and Figure 7.3 respectively show the percentage values of precipitation for a month and for the accumulated precipitation for the last three months.

Maps are then drawn up with values of the corresponding indicators. These data are generated by the River Basin Authorities and are sent periodically to CEDEX where a common database is kept.

7.6 The Júcar River Basin District

The Júcar River Basin District (Júcar RBD) is located on the eastern part of Spain (Figure 7.4). It is made of a group of different river basins and covers an area of 42,989 km^2. From the 17 autonomous communities in the Spanish territory, the Júcar RBD encompasses part of four of them: Valencia, Castilla-La Mancha, Aragón, and Cataluña, just including a small area from the latter.

The population within the district is about 4,360,000 inhabitants (2001), which means that about 1 in every 10 Spaniards lives in the Júcar RBD. In addition to this number about 1,400,000 equivalent inhabitants are added

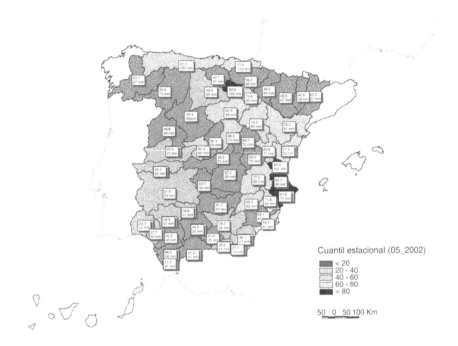

Figure 7.2 Precipitation percentages for a month (May 2002).

Figure 7.3 Accumulated precipitation percentages for the last 3 months (May 2002).

Figure 7.4 Territorial area of the Júcar River Basin Authority.

due to the tourism, primarily in the Valencia community. Nevertheless, the Júcar RBD is a district of great contrast since population density ranges from over 20,000 inhabitants per square kilometer in the metropolitan area of the city of Valencia at the coast, to less than two inhabitants per square kilometer in the mountainous areas of the province of Cuenca at the western part of the district.

The area has a Mediterranean climate, with an average annual precipitation of 504 mm (MIMAM, 2000b), varying from 250 mm in the south to about 800 mm in the north of the area (Figure 7.5). This situation necessitates defining different levels of regional vulnerability to droughts. The precipitation over the basin produces a mean annual runoff of 80 mm, which represents approximately 16% of the precipitation. Renewable water resources are about 3400 hm³/year (MIMAM, 2000b).

The amount of 504 mm/year corresponds to a volume of 21,220 hm³/year over the land surface of the territory. About 85% of this precipitation is consumed through evaporation and transpiration by the soil-vegetation complex. The remaining 15% comprises the annual runoff of 3250 hm³/year (Figure 7.6).

An analysis of the mean annual precipitation (Figure 7.5) in the Júcar river basin district for the 1940/1941–2000/01 period allows differentiating periods

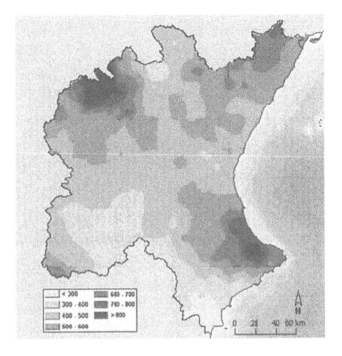

Figure 7.5 Mean annual precipitation (mm) in the Júcar River Basin area.

according to their behavior, with the most important being the humid periods of 1958–1977 and 1986–1990, and the driest periods of 1978–1985, 1991–1995, and 1997–2000 as is shown in the deviation graph in Figure 7.8.

The Júcar RBD is characterized by long drought periods, in some cases reaching even 10 years. An index that reflects the annual deviation from the mean annual rainfall is the Standard Precipitation Index (SPI), shown in

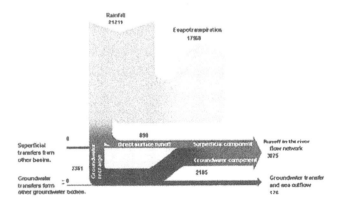

Figure 7.6 Water cycle in natural regime for the Júcar RBD (figures in millions of m³).

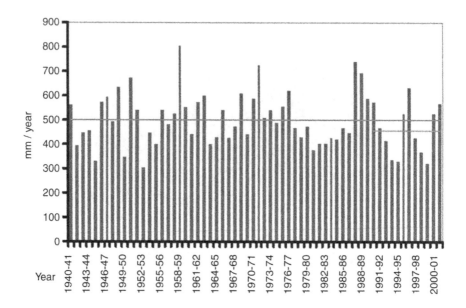

Figure 7.7 Yearly rainfall in the Júcar River Basin District.

Figure 7.8 Rainfall unit deviation graph for Júcar River Basin District.

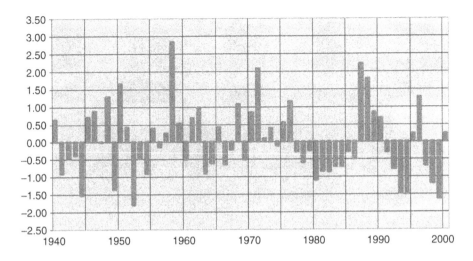

Figure 7.9 SPI values for annual precipitations in the Júcar RBD.

Figure 7.9, which is a normalized index used for quantifying deficits in the volume of precipitation for any given period of time.

The spatial deviation maps for the years corresponding to the 1977–1986 and 1991–1995 drought periods are shown in Figure 7.10. These maps represent, for each year, the percentage of variation of the annual precipitation with respect to the mean annual values corresponding to the period 1940–2000. The bars in Figure 7.9 show the highest percentage variation from the period mean value (1940–2000), which indicates that those are the driest years for the represented drought period.

Within the Júcar River Basin District the water resources used come from superficial and ground water origins. Superficial water resources have been used historically since Roman and Arab times. Nowadays, these resources are being regulated through large dams (Figure 7.11). The reservoir capacity for the whole basin is of 3300 hm³; of high importance are the reservoirs of Alarcón, Contreras, and Tous in the Júcar river, and Benageber in the Turia river. The resources coming from ground water, with a value of 2500 hm³/year, represent slightly more than 70% of the total resources used, which reflects the importance of this type of resource in the basin (MIMAM, 2000b).

The joint use of surface water and ground water is quite common within the basin, with clear examples being the Plana of Castellón, La Marina Baja, or the Ribera of the Júcar. However, the intensive use of ground water has produced overexploitation problems in some of the hydrogeological units, such as the ones of the exploitation system Vinalopó-Alacantí, the ones from coastal plateaus of the province of Castellón, or the hydrogeological unit of the Mancha Oriental aquifer.

Regarding the reuse of nonconventional resources, it is important to mention the high potential of reuse (treated wastewaters), which represents

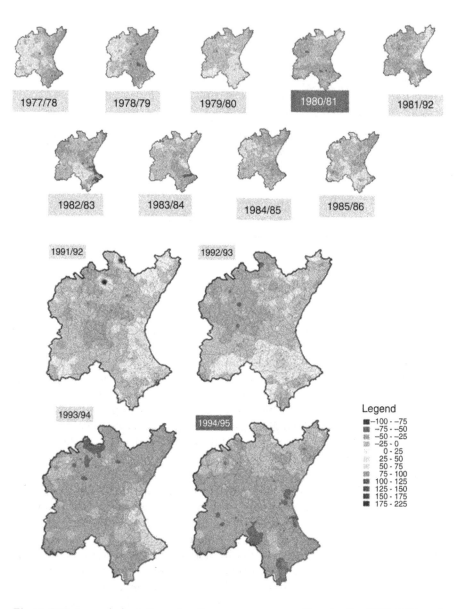

Figure 7.10 Annual deviations for the years corresponding to the 1977–1986 and 1991–1995 drought.

one of the highest achievements in Spain. The total water demand in the basin is 2962 hm³/year, being distributed into sectors as 563 hm³/year for urban use, 2284 hm³/year for agricultural use, 80 hm³/year for industrial use, and 35 hm³/year for refrigerating energy plants, with the highest percentage being the one corresponding to agricultural use, which represents 80% of the total demand (MIMAM, 2000b).

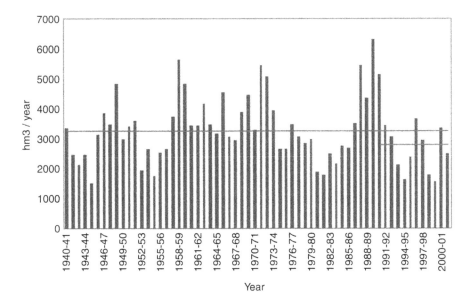

Figure 7.11 Annual runoff in the Júcar River Basin District.

In general, the territorial area of the Júcar is characterized by having a balanced equilibrium between renewable resources and water demands (CHJ, 1999), although water shortages occur in some areas, especially in the ones located in the coastal strip of the province of Castellón, in the Mancha Oriental aquifer, and in the exploitation systems of Vinalopó-Alicantí and Marina Baja.

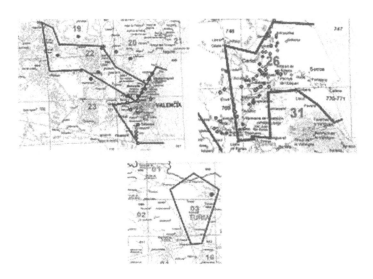

Figure 7.12 Emergency wells present during the 1991–1995 drought and aquifers affected.

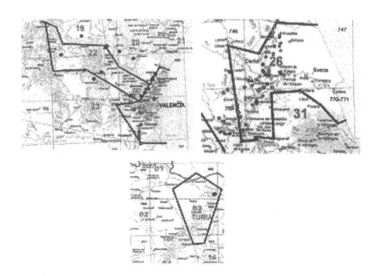

Figure 7.13 Drought indicator system in the Júcar River Basin District.

7.6.1   Recent droughts occurred in the Júcar River Basin

The most intense droughts recently suffered in the Júcar river basin occurred during the period of 1991–1995. The shortage on surface water resources made the Ministry of Environment declare an emergency of the development of works for ground water abstraction in the following areas: the public water supply of the town of Teruel and the agricultural traditional irrigation systems of "Acequia real del Júcar," "Ribera Alta" in Júcar river, and "Vega de Valencia" in Turia river (see Figure 7.13).

Table 7.1 shows a summary of those works developed by the General Directorate of Hydraulic Works and the Júcar River Basin Authority, which indicates users affected, the number of pumping wells, and flows.

7.6.2   Drought indicators in the Júcar River Basin District

A specific procedure has been developed in the Júcar river basin for follow-ups of droughts based on a system of indicators of hydrological variables (flow

Table 7.1 Emergency Drought Actions Based on Ground Water

Use	Users affected	Pumping wells	Flow (l/s)
Urban	Teruel city	4	280
Agricultural	Channel "Acequia real del Júcar"	43	3367
Agricultural	Júcar "Ribera Alta" area	7	629
Agricultural	Turia "Vega de Valencia" area	6	495

river gauges, aquifer water levels, water storage at reservoirs, river flow gauging, etc.), representative of the hydrological situation of each of the exploitation systems defined in the Hydrological Júcar River Basin Plan. Quarterly reports are made and are available for public use from their website (http://www.chj.es).

The different phases of this methodology are:

1. Identification of water resource areas (origin) associated with specific demand units (destination)
2. Selection of the most representative indicator for the evolution of water resources for each of the previously identified areas
3. Compilation of hydrological temporal series associated to each of the previously selected indicators
4. Establishment of specific weights for the different indicators
5. Continuous follow-up of hydrological series associated to indicators, and elaboration of the corresponding periodical reports

Depending on the type of variable, a corresponding timing for follow-up and a specific processing is done. For instance, for the pluviometric data a year is considered as a representative time period, three months for superficial stream gauging, and for stored volumes, the last measure taken before issuing the report, which corresponds to a month. These previous indicators are not directly comparable; therefore, a nondimensional status index has been defined, which allows establishing spatial and temporal comparisons. This status indicator has been defined taking into account:

- The mean is the simplest and strongest statistic unit; therefore, it must have an important weight in the definition of the status indicator, as it is reflected in the formulas applied (Figure 7.14).

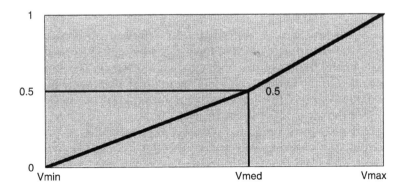

Figure 7.14 Nondimensional status indicator.

- In order to standardize the indicators and be able to give them a nondimension numerical value, a formula has been adopted (in which the status indicator $[I_e]$ is defined with values that range from 0, corresponding to historical minimum values, to 1, corresponding to the maximum historical value, according to the following expressions:

Status indicator

$$\text{If} \quad V_i \geq V_{med} \Rightarrow I_e = \frac{1}{2}\left[1 + \frac{V_i - V_{med}}{V_{max} - V_{med}}\right]$$

$$\text{If} \quad V_i < V_{med} \Rightarrow I_e = \frac{V_i - V_{min}}{2(V_{med} - V_{min})}$$

I_e Status indicator
V_i Measured mean value for the analyzed period
V_{med} Mean value for the historical period
V_{max} Maximum value for the historical period
V_{min} Minimum value for the historical period

If the measured value ranges between the mean and the maximum value, the status indicator will give a result between 0.5 and 1, whereas if the measured value is lower than the mean value, the result will be between 0 and 0.5.

The following four levels are used to characterize a drought situation, which are graphically represented in Figure 7.15

$I_e > 0.5$ Green level (stable situation)
$0.5 \geq I_e > 0{,}3$ Yellow level (pre-alert situation)
$0.3 \geq I_e > 0{,}15$ Orange level (alert situation)
$0.15 \geq I_e$ Red level (emergency situation)

The stable situation is associated with a better hydrological situation than the mean situation; the rest of the levels are established to differentiate situations below the mean one and are useful to launch the different measures detailed in the Drought Special Plan in order to mitigate the effects of the droughts. Figure 7.16 shows the temporal progress of the global status indicator of the Júcar River Basin District.

From the experience acquired since the implementation of this indicator and from quarterly reports, it is derived that this indicator is a versatile tool of analysis, and even though it presents limitations since it is considered a discrete estimator, it allows a quick examination of the hydrological resources

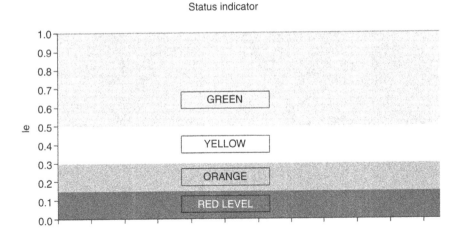

Figure 7.15 Status indicator adopted in the Júcar River Basin.

status in the whole basin area, as well as a description of the temporal evolution of the hydrological status.

The drought situation affects different areas at different times as it is shown in Figure 7.17, which shows the mean weighted values of the status

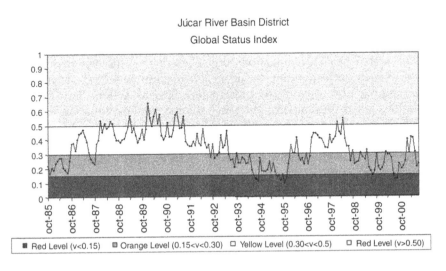

Figure 7.16 Temporal evolution of the global status indicator of the Júcar River Basin District.

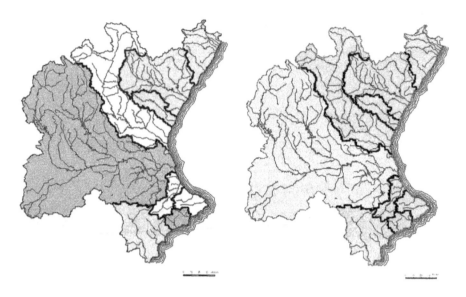

Figure 7.17 Status indicator for the different water resource systems for April 2002 (left) and for June 2002 (right).

indicator for each of the water resource systems of the Júcar River Basin District.

7.6.3 *The Júcar River Basin Drought Special Plan*

The development of the Drought Special Plan started in 2002 to enforce Article 27 of the National Hydrological Plan previously mentioned, which indicates that basin organizations will elaborate within the hydrological plans of each corresponding basin, in the maximum period of two years since coming into force of the present law, Special Action Plans for Alert Situations and Eventual Droughts, including system exploitation rules and measures to apply to the hydraulic public domain use.

The main objectives of the Júcar River Basin Drought Special Plan are to anticipate droughts and to foresee solutions to satisfy demand, avoiding situations of undersupply.

The bases for the Drought Special Plan are:

- Present indicators that provide a quick drought status early enough to act according to the forecasts of the plan
- Provide knowledge of the resources system and its elements' capability to be strained during scarcity situations
- Provide knowledge of the demand system and its vulnerability toward droughts, organized by priority degrees

- Present structural and nonstructural alternatives to reduce drought impacts and adaptation according to the status indicator
- Measure the cost of the implementation of measures
- Adapt the administrative structure for its follow-up and coordination among the different administrations involved
- Develop a public information plan and a plan for the staff in charge of water supply systems

The development of the Drought Special Plan was to be completed in 2004.

The main mitigation measures included in the Drought Special Plan can be grouped into different categories: structural measures (new pumping wells, new pipes, use of new desalination plants, etc.), and nonstructural measures (water savings by applying restrictions to the users, increase in the use of ground water, etc.). Next the different measures proposed are described in more detail.

- Exploitation rules for drought situations: Simulation models are used to study the exploitation rules and constraints for water demands for reaching an optimal drought management. An example is the Júcar model developed for the Júcar River Basin Authority by the Polytechnic University of Valencia (see Figure 7.18).
- Aquifers of strategic reserve: The use of ground water allows increasing the supply of resources to satisfy water demands. This use presents the advantage of not needing large infrastructures for its exploitation, and, in addition, if the aquifers are located in an area that shows deficit, large transportation works are not necessary. It is important that aquifers are considered as a strategic reserve.
- Temporary exploitation of ground water reserves: Some aquifers can be temporarily exploited above their renewable water resources levels without causing severe environmental problems and later cease their exploitation to allow their regeneration. The exploitation of alluvials allows flows to be readily available, though its temporal sustainability is limited.
- Emergency wells: During a crisis situation, emergency wells can be built to exploit local ground water resources, which are not traditionally used (Figure 7.19). As mentioned earlier, in 1995 a significant number of emergency wells were built in the Júcar River Basin to use groundwater resources, in order to avoid adverse conditions caused by droughts (see Figure 7.12).
- Desalination: In a drought situation it is possible to intensify the use of desalination water resources, especially with mobile desalination plants, which can be transported from one area to another.

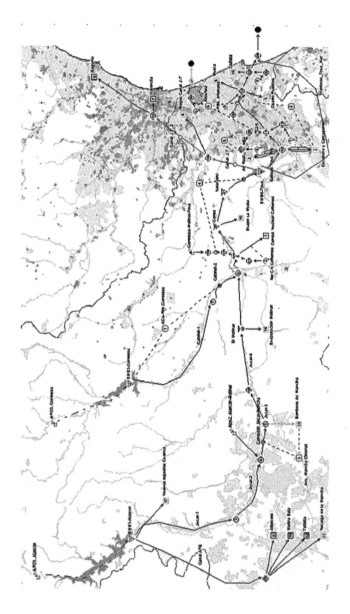

Figure 7.18 Júcar simulation model.

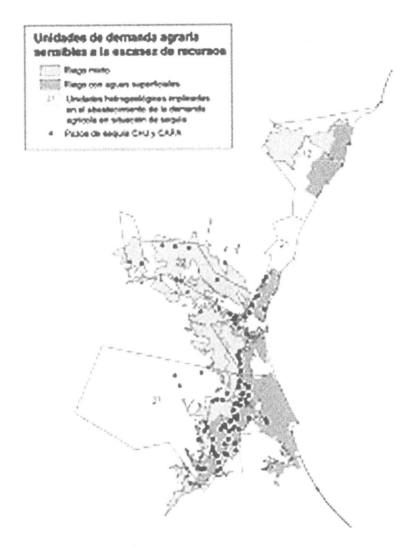

Figure 7.19 Emergency wells in the Júcar RBD coastal areas.

The Drought Special Plan must activate and serve as a reference framework for water supplying emergency plans for towns over 20,000 inhabitants (Figure 7.20), and must:

- Define a proposal of acting measures on supply and demand
- Define prioritizing and layering measures: supplying alternatives, changes in the management system, demand restrictions, decrease of environmental conditions restrictions
- Adapt and evaluate the different measure types within the different exploitation systems and demand units

Figure 7.20 Activation emergency plans (municipalities/systems > 20,000 inhabitants).

- Present a plan of measures, with progressive introduction, according to the status indicators: drought status, public administration actions, actions issued to the user, sanctions
- Provide coordination rules among administrations

7.7 Conclusion

Article 1e of the Water Framework Directive establishes as one of its objectives contributing to alleviating drought effects. The WFD establishes in Article 4.6 that precise extreme situations might be temporarily exempt of complying with the defined objectives. However, it must be specified in the Basin Hydrological Plan the conditions in which these extreme situations can be declared, including an appropriate indicators' system.

The legal Spanish system establishes a double-way action:

- The Water Law establishes an emergency action against drought situations with a focus on crisis situation.
- The National Hydrological Plan defines a planning focus, indicating the necessity of designing a global system of hydrological indicators

that allows foreseeing these situations, as well as the requirement for elaborating a Special Plan and an Emergency Plan, being responsible for the first plan the Spanish Basin Organisations, and for the second type of plan, public administrations in charge of supplying systems that cover towns having 20,000 inhabitants or more.

The first approach has been traditionally used in Spain in past drought situations, with a series of actions heading toward increasing water resources by developing hydraulic works, especially ground water abstractions, which allows benefiting from the high storage capacity of the aquifers.

The second approach tends to design and establish an indicator system that allows foreseeing extreme situations, establishing levels or thresholds depending upon the degree of the drought, and consequently developing a series of actions aiming to delay or impede critical situations. The final intention is to include droughts as one more situation to take into account when planning and managing water resources.

References

CHJ. (1999). *Plan Hidrológico de la cuenca del Júcar.* Ministerio de Medio Ambiente. Confederación Hidrográfica del Júcar. Depósito Legal V-3211-1999.

Estrela, T., Marcuello, C., and Dimas, M. (2000, October). *Las aguas continentales en los países mediterráneos de la Unión Europea.* Ministerio de Fomento y Ministerio de Medio Ambiente.

MIMAM. (2000a, September). *Delimitación y asignación de recursos en acuíferos compartidos. Documentación técnica del Plan Hidrológico Nacional.* Madrid: MIMAM.

MIMAM. (2000b). *Libro Blanco del Agua en España.* Madrid: MIMAM.

EC. (2003, January). Common Implementation Strategy for the Water Framework Directive (2000/60/EC). Identification of water bodies.

EC. (2002a, December). Common Implementation Strategy for the Water Framework Directive (2000/60/EC). Working Group GIS, Guidance Document on Implementing the GIS Elements of the WFD.

EC. (2002b, August). Common Implementation Strategy for the Water Framework Directive (2000/60/EC). Best Practices in River Basin Management Planning. WP1 Identification of River Basin Districts in Member States. Overview, criteria and current state of play.

chapter eight

Water reuse and desalination at Comunitat-Valenciana region, Spain

Vicente Serrano-Orts
Urbanisme i Transports, Valencia, Spain

Javier Paredes
Universidad Politécnica de Valencia, Spain

Joaquín Andreu
Universidad Politécnica de Valencia, Spain

Contents

8.1 Introduction

According to the territorial organization of the Spanish state, Comunitat-Valenciana is one of the 17 autonomous regions into which Spain is divided. This chapter gives an updated vision of the wastewater direct reuse and desalination activities in the region. The chapter is split into three parts. The first part explains the legislative, territorial, and social frameworks, as well as the water situation in the region. The second part gives a short summary of the water reuse in Spain and a more detailed description of the reuse in Comunitat-Valenciana with current and future actions. The third part describes, in a similar way, the water desalination activities.

8.2 Legislative, territorial, and social framework

Comunitat-Valenciana is located east of the Iberian Peninsula in the Mediterranean coast. Its climate is typically Mediterranean with mild winters and warm summers. It represents 4.6% of the total Spanish area, and it has a population of approximately four million people, 10% of the total population. Therefore, it is a densely populated region in relation to the rest of Spain. The activities of the region generate 10% of the national GNP.

In Comunitat-Valenciana, 738,000 h are cultivated, and more than 50% of the land is irrigated. Sixty-six percent of the total irrigated crop production is exported, and this represents a third of the Spanish agrarian export and 3.3% of the GNP of the region. The most important crops are citrus, fruit trees, and vegetables. This sector uses approximately 80% of all of Spain's water resources, and the most water scarce areas are the most profitable economically within the agrarian sector.

Industry generates 28.1% of the GNP of Comunitat-Valenciana and is characterized by its diversity. The services sector is important in terms of GNP (68.6%). Due to the conjuncture of different factors, such as the Mediterranean weather and its coastal situation, tourism is one of the main activities. It produces high seasonal population increases along the Valencia littoral (more than 20 million persons per year, and 45% of them are foreigners).

The Spanish legislature is divided into three levels: central government, autonomous regional governments, and municipalities. Water resources management is under the central government's responsibility if the basin extends to different autonomous regions and under the autonomous regional government if the whole basin stands within its geographical limits. Most environmental competencies are managed by the autonomous regional government, whereas water supply and wastewater treatment are under municipalities control. The authority for financial and technical aids, given to the municipalities for the water infrastructure performance, has been given to the autonomous government (RD 1871/85). To compensate for this the government of Comunitat-Valenciana created the wastewater entity Entidad de Saneamiento through the 2/92 Law. This entity works as a public enterprise responsible for wastewater treatment.

8.3 Water situation of Comunitat-Valenciana

In terms of water resources, most of Comunitat-Valenciana territory is within the Júcar Basin Agency. The rest is included within the Ebro basin (3.5%) or the Segura basin (5.2%). The region has four main rivers — Júcar, Segura, Turia, and Mijares — that together represent 80% of the regional water resources. Furthermore, other rivers of lesser flow complete the region's water resources. Most have a torrential regime in which strong floods alternate with long dry periods. This is more acute in some areas of the south due to their very low rainfall, usually less than 200 mm/year.

The volume of water available in Comunitat-Valenciana is about 2681 hm^3/year, where 48% is superficial water and 52% ground water. There are 431 hm^3/year of return water, 170 hm^3/year provided by external basins, and 185 hm^3/year coming from wastewater and desalination efforts. Therefore, the total resources available are 3476 hm^3/year. The appraised water demand, including environmental requests, is approximately 3667 hm^3/year. That indicates a global shortage of 191 hm^3/year. This deficit affects mainly the agricultural sector, leading to practices that seriously affect the environment: overexploitation of aquifers, irrigation with poor quality water, or insufficient irrigation.

The water stress situation is particularly important in the subsystem of the southeast, where the demand (400 hm^3/year) is 157 hm^3/year or higher than the water availability. That is also the case in the Vinalopó-Alacantí region.

To overcome these situations of scarcity, different actions are being taken, leading in three directions: (a) the reorganization of the internal resources distribution by sending the water from systems with surplus to the most deficit ones; (b) the control of the demand, in order to save water especially through modernization of the irrigation structure (substitution of flood irrigation to localized irrigation systems), and (c) more intensive use of non-conventional resources (reuse of reclaimed wastewater and desalinated water).

But these measures are limited and are not enough to equilibrate the existent water shortage and to ensure the short- and medium-term stability. This is why it is considered necessary to import water from basins outside Comunitat-Valenciana limits.

Therefore, the reuse of reclaimed wastewater is gaining great importance in this area. There are subsystems where wastewater reuse is a significant contribution. As a result of the need to reuse water, large investments have been made for treatment plants and for improving the quality of reclaimed wastewater.

Water quality is a problem in many parts of Comunitat-Valenciana, and it becomes especially serious in the areas where resources are scarce. With regard to the ground water, both agriculture and urban pressures exist in the coastal areas, causing overexploitation of aquifers, seawater intrusion problems, diffused pollution problems by nitrates, and industrial located pollution.

8.4   Wastewater reuse

8.4.1   Wastewater reuse in Spain

The volume of wastewater directly reused in Spain is about 230 hm³/year (MIMAM, 2000), and more than 50% takes place at Comunitat-Valenciana. Eighty-nine percent of the total volume of reused water is used for agriculture and the rest on golf course irrigation, municipal uses, industrial reprocessing, and also on environmental requirements (the hydrologic plans of the Júcar and Segura basins).

Among the rest of the autonomous regions, the Canary islands (17.5 hm³/year) and the Balearic islands (15 hm³/year) are especially important, with golf course irrigation being the main application. Andalucia does both agricultural and golf course irrigation. There are other applications of interest in Spain such as environmental regeneration of the Manzanares river (Madrid) with reclaimed wastewater, which is also used for irrigation of parks and golf courses. In Catalonia, the Costa Brava Consortium is recognized because of its extensive irrigation of the golf courses as well as its use of purified residual water for a theme park. Finally, of special interest is the case of Vitoria-Gasteiz, where a very advanced tertiary system allows the use of reused water for irrigation of 4000 h of land.

8.4.2   Reuse at Comunitat-Valenciana

Direct reuse has been practiced since the middle ages at Comunitat-Valenciana. At that time, drainage facilities were combined with the irrigation facilities in the lower part of the Segura basin. The aim was to collect the water surplus for reuse on the fields downstream.

At present, 125 hm³/year of reclaimed wastewater is being reused at Comunitat-Valenciana, mainly in agriculture, golf courses, and municipal and environmental uses. In the mid-1980s some competencies were transferred from the central government to the autonomic one. The aim of these competencies was to provide technical and financial support for the residual water treatment substructures. The autonomous government developed the legal framework for urban drainage and treatment facilities, mentioned before, and started a Sanitary Plan that is now in place. As a result of that plan, there are now 375 treatment plants that treat 450 hm³/year of wastewater. These plants cover practically all of the total population's needs and have an average performance close to 90% in reducing the organic matter in the water.

More recently the Second Sanitary Plan has been approved. Within this plan the reuse of reclaimed wastewater has gained relevant importance: the installations where water must be reused have been identified as well as the additional treatments that must be applied to the process.

The amendments of the Spanish Water Law of 1999 added some simplifications of the general process of concessions of treated wastewater (simple authorization for a farm water user and a contract between the reuse concession holder and the dumping concession holder). Nevertheless, in Spain, more legislature is still pending regarding the regulation of some aspects such as the dumping and reuse holders responsibilities and the testing and monitoring required.

In some areas of Spain where reuse is necessary, such as Comunitat-Valenciana, the use of treated water is taking place through an agreement between the treatment plant holder and the user who receives or transports the reclaimed water to the irrigation ditch or to the storage facility.

The problems found on the resource regulation of any water exploitation exist also for the regulation of the reuse of reclaimed water: the water production is practically constant every day, but the demand, especially in agriculture, is irregular. This situation necessitates the use of pools where tracking of water quality is done, especially to control algae development and to examine bacteriologic evolution.

The lack of regulations concerning water reuse also affects the quality standards of purified water. The irrigated crops in Comunitat-Valenciana are mainly citrus and vegetables, where it has seen a progressive transformation from traditional furrow irrigation to localized irrigation systems. The Second Sanitary Plan expects, besides the use of treated water on crop irrigation, the use of reused water for golf course irrigation and industrial reprocessing. This is why the quality standards established by the EPA have been chosen for water reuse in public parks, golf courses, and for irrigating fresh consumer crops. The current limits imposed are: DBO5 < 10 mg/l, turbidity < 2 NTU and no detectable coliforms. The tertiary treatment applied to reach these limits consists of filtration (usually after a decantation) and then disinfection by UV rays.

Among the most interesting actions are the following:

- The reclaimed wastewater of Alacantí is pumped 400 m to the Medio Vinalopo area. This water is used on the vineyards and vegetable crops and comes from the treatment plant of Rincon de Leon, which has a production capacity of 60,000 hm^3/year. The net water distribution of the purified water includes links to pools used in order to optimize purified water application.
- Marina Baja action. This region has shown tourist explosion within the past 30 years, and both permanent and seasonal population are still growing. To meet this increasing urban demand the inland irrigators have transferred their water rights to the coastal city's benefits, provided that the purified residual water would return to them for reuse.
- Valencia metropolitan area. Reclaimed wastewater of Valencia is being used by the traditional irrigated canals. Nowadays, there are five main treatment plants, and water is pumped to the different irrigation ditches used for Valencia's fertile irrigated area. The most important plant is the Pinedo Plant, which has a water treatment capacity of 350 hm^3/year that is used both for irrigation and for Albufera lake environmental regeneration.
- The existing plant of Castellon de la Plana has been enlarged, and a tertiary treatment has been set up, with a capacity of 45 hm^3/year.

The water produced is transported to a pool close to the treatment plant where the water is at the disposal of the irrigators.

Environmental uses of reclaimed wastewater is another use, separate from the traditional agricultural use. In this case, reclaimed wastewater is used to maintain ecological flows and wetlands conservation (Albufera, Delta del Millars, Hondo de Elche, Clot de Galvany). There are also planned industrial actions, such as irrigation of green areas and streets cleaning.

The whole volume of direct reuse expected with the Second Sanitary Plan at the Comunitat-Valenciana is about 270 hm³/year, including the periods where no irrigation is conducted, with nearly 60% reclaimed water, which is the real limit of reuse. Thus, 10% of the water resources assigned to irrigation in the region will come from reclaimed wastewater.

8.5   Desalination

8.5.1   Desalination in Spain

Spain is one of six countries with the highest production of desalinated water of the world, and the first one in Europe. There are 700 desalination plants, which produce 220 hm³/year. Generally, the plants for desalination of brackish water are more numerous and have a lower capacity than the seawater desalination plants (Medina, 2001). In volume, the desalination of marine water represents 93 hm³/year; whereas brackish water represents 127 hm³/year. Seventy-two percent of the desalinated water is used on urban or industrial development and the remaining part is used for agricultural irrigation. However, this 28% indicates that Spain is one of the countries with the highest volume of desalinated water devoted to agriculture in the world.

Historically, the investment on desalinating plants has been the result of periods of drought that have taken place in this country. The highest volume of desalination takes place in the Canary and Balearic islands, basically to allow the development of tourist activities in both communities. Water has even been carried by ship for human supply in Mallorca.

The first desalination plant was built in 1964 in Lanzarote and produces 2500 m³/day. Several islands such as Lanzarote and Fuerteventura do not possess alternative resources, and the consumption of each inhabitant there is the lowest in Spain (80 l/person/day). In Gran Canaria, 90% of the population is supplied by desalinated water (Rico et al., 1998). The three larger plants in Spain are Las Palmas III (58,000 m³/day); Bahia de Palma (53,000 m³/day); and Costa del Sol Occidental (45,000 m³/day). Economically, the development of the technology has reduced the threshold from 0.6 euro/m³ in larger plants (over 50,000 m³/day) (MIMAM, 2000).

The other autonomous communities in which desalination is carried out as a nonconventional source of water resources are, once again, those situated on the east and southeast of the peninsula, because of the desalinated volume, Comunitat-Valenciana, Murcia, and Andalucia.

8.5.2 Desalination at Comunitat-Valenciana

The overexploitation of aquifers in many coastal areas has increased the application of desalination techniques to brackish or marine water in many places in the Spanish southeast and also in Comunitat-Valenciana. The number of brackish water plants is much higher than that of marine water plants. The origin of many of them has been the salination of aquifers due to overexploitation problems. The irrigation of the campus of the University of Alacantí with desalinated water, taken from a well drilled within the campus, is a very interesting example of this.

According to the origin of the water, the desalination activity in Comunitat-Valenciana can be divided into three categories: the desalination of brackish water, the desalination of marine water, and the desalination of reclaimed wastewater.

For obvious economic reasons, the first desalination plants with an agricultural destination took advantage of brackish ground water resources. The current price, the cost of reverse osmosis desalination plants of medium size (between 5000 and 10,000 m^3/day) for brackish water (concentrations of 5000 μS/cm) versus marine water plants, is 50% lower. But environmental problems, which are basically due to dumping of the brine, have caused the desalination concentrates to seep into the marine water.

Osmosis is also used to desalinate superficial brackish water. In this particular case, the treatment before the membrane acquires special importance so as to make this treatment really effective.

Reverse osmosis is usually employed for human consumption water in Comunitat-Valenciana. The water intake is from wells placed next to the coastline and, consequently, the treatment develops through double filtration on sand and through cartridges. The remineralization is accomplished by adding salts or by mixing it in tanks with resources from another source used for its supply.

The practice of dumping the brine into the sea, utilized in different facilities currently operating, has been studied. The most used system is direct dumping by choosing the area of the coast with enough dilution and renewal. This alternative has been chosen instead of dumping away from the coast through submarine outlet due to the effects within the Posidonia population.

Very relevant examples of sea desalination plants used to produce water for human consumption, according to the treatment plan explained before, are cited below:

- In the region of Marina Alta, the plants placed in Teulada-Benitachell with a capacity of 10,000 m^3/day and the one found in Javea with a capacity of 26,000 m^3/day.
- In the region of l'Alacantí, the plant of the Alacantí channel with a capacity of 55,000 m^3/day.

The process for desalination of sewage treatment tries to eliminate the infiltration of the phreatic level from the coastal towns (drains and general sewers), and this process is being put into practice at the sewage treatment plants of Alacantí (Rincon de Leon) and Benidorm. In both cases, the concentration of the water input is about 2500 µS/cm, while the salinity of the water produced is set according to the needs of the users. The needs fluctuate between 1000 µS/cm at the plant of Benidorm and 600 µS/cm at the plant of Rincon de Leon, according to the possibility of mixing it with nondesalinated water volumes.

In both cases, reverse osmosis has been chosen as the system for desalination, preceded by a pretreatment with microfiltering or ultrafiltering membranes. Because of the magnitude of the volumes that have to be desalinated (25,000 m³/day, in both cases), it was decided to carry out a test on a natural scale in order to choose the most adequate pretreatment system, according to the water that is really going to be desalinated and that comes, in both cases, from a secondary treatment.

The bidding carried out has highlighted that the osmosis system and the pretreatment with membranes is cheaper than the reversible electrodialysis and that the cost of the pretreatment membranes is higher than that of desalinated membranes.

8.6 Conclusion

It is difficult to summarize in a few pages the work that has taken place in the past seven years on direct reuse and desalination in Comunitat-Valenciana. However, it is important to outline that in the areas where water is a limited good, a huge effort is being made to make the most efficient use of the resource. On the other hand, it should be highlighted that this effort is being made because in this area agriculture is profitable enough so as to assume the costs of desalinated or reused water and still obtain benefits.

As far as the human supply is concerned, in those exploitation systems that are particularly short of water, the necessary resources are being obtained through the desalination of marine water through membranes by reverse osmosis. In these systems, it is also common to obtain additional resources through an exchange of water rights with the agricultural users, who in turn receive reclaimed wastewater.

8.7 Acknowledgment

Thanks are due to Conselleria de Obras Publicas de la Generalitat Valenciana for the information supporting the chapter. The writing and editing have been funded by the contracts WAMME: Water Resources Management Under Drought Conditions (within the European INCO.MED Program), and SEDEMED: Secheresse and desertification dans le bassin Mediterranee (within the European INTERREG program).

References

Medina, J. A. (2001, June). La desalación en España. Situación actual y previsiones. Conferencia Internacional: El Plan Hidrológico Nacional y la Gestión Sostenible del Agua. Aspectos medioambientales, reutilización y desalación, Zaragoza.

MIMAM. (2000). Libro Blanco del Agua en España. Secretaría de Estado de Agua y Costas. Dirección general de obras Hidráulicas y calidad de Aguas. Ministerio de Medio Ambiente.

Rico, A. M., Olcina, J., Paños, V., and Baños, C. (1998). *Depuración, desalación y reutilización de aguas en España*. Editorial Oikos-Tau.

chapter nine

Role of decision support system and multicriteria methods for the assessment of drought mitigation measures

Giuseppe Rossi, A. Cancelliere, and G. Giuliano
University of Catania, Italy

Contents

9.1 Introduction

Today the dramatic situation of several countries, experiencing an increasing risk of water scarcity and quality degradation of water resources, needs an urgent application of the new ethic concept of sustainable development (Rossi, 1996). The application of general principles of sustainability, as suggested by the Brundtland Commission (WCED, 1987), to the water sector requires significant changes or reconsideration in a new context of traditional methodologies generally applied for decision-making processes in all phases of planning, design, and operation of water systems (Simonovic, 1996). As an example, the identification of the objectives of the water resources development should include the new concepts of environmental integrity and social equity besides technical performance and economic efficiency, already accounted for in an integrated water resources management. Further, the time line for assessing consequences of the projects should be chosen according to the needs of future generations. The procedure for implementation of actions should be revised to promote stakeholders participation in the decision-making process and to balance the market mechanism with democratic control.

Among the several tools that can contribute to improve such an integrated and sustainable management of water resources, decision support systems (DSS) play a central role, since they enable decision makers to better understand the problem at hand, to explore alternative courses of action, and to predict consequences with the necessary detail in order to choose the preferable solution (Loucks and Da Costa, 1991; Andreu et al., 1996; Reitsma et al., 1996; Simonovic, 1996; Andreu et al., 2001). The term DSS often refers to different types of computer-based modeling tools. Here the adopted definition assumes that a DSS includes information systems oriented to store, retrieve, and modify data required for analyzing the water resources systems; mathematical models able to analyze the systems behavior and to evaluate consequences; and user interfaces to facilitate the input and output of data, as well as to improve the interpretation of results. Once the consequences of each course of action has been determined, selection of the preferable alternative can be performed by multicriteria analysis (MCA), which is generally recognized as one of the most convenient methodologies to compare alternatives exhibiting complex multidimensional impacts, taking into account the different levels of the preference of the various groups of interest (Goicoechea et al., 1982).

This chapter presents an application of DSS and MCA to a very important issue in the framework of water resources management, namely the identification and assessment of drought mitigation measures. After a brief illustration of a proactive approach for mitigating drought impacts, a procedure aimed at evaluating drought vulnerability of water supply systems, identifying long- and short-term measures for drought mitigation and comparing and ranking the measures, is presented. The role of MCA is also discussed with particular emphasis to a technique developed in a sustainable development context (Munda, 1995). Special focus is placed on the application of the proposed methodology to three case studies located in Italy (southern Sicily and Sardinia) and Spain (Valencia region).

The procedure has been developed within Work Package 7 of the European Project WAMME, Water Resources Management Under Drought Conditions: Criteria and Tools for Conjunctive Use of Conventional and Marginal Waters in Mediterranean Regions.

9.2 Assessment of drought mitigation measures through a multicriteria approach

9.2.1 Measures for coping with drought

Drought is a complex hydrometeorological phenomenon, originated by meteorological anomalies that reduce precipitation, but strongly affected by the state of the various components of the hydrologic cycle (Wilhite, 2000). In spite of its basic nature of natural hazard, drought can also be considered a man-affected phenomenon (Rossi, 2000). This fact derives, first of all, from the consideration that drought is perceived as an adverse phenomenon only where a human community exists; moreover drought impacts can largely differ according to the level of withdrawals with respect to the available water resources as well as to the structural measures and policies adopted to manage hydrological variability (e.g., stream flow regulation through reservoirs or long-distance uni- or bi-directional water transfers). In particular, a drought of fixed duration and severity could produce a wide range of consequences according to the vulnerability of the water system and to the drought preparedness or mitigation strategies (Cancelliere et al., 1998). Although this in principle should facilitate the mitigation of most severe impacts of drought by implementing appropriate measures, in practice, improving drought preparedness has received very limited attention by technical, management, and political sectors almost everywhere except in a few countries.

The measures to be implemented to improve drought preparedness and to mitigate drought impacts can be classified in several ways. A first classification (Yevjevich et al., 1983) refers to purposes of the measures, distinguishing them into three main categories:

1. Water supply increase oriented measures
2. Water demand reduction oriented measures
3. Drought impact minimization measures

The first two categories of measures aim to reduce the risk of water shortage due to a drought event by modifying supply or demand, while the third category is oriented to minimize the environmental, economic, and social impacts of drought.

A second classification focuses on the type of response to drought events, distinguishing between a reactive and a proactive approach (Rossi, 2000). The *reactive approach* consists of measures adopted once a drought occurs and its impacts are perceived. It includes the measures taken during and after the drought period to minimize the impacts of the drought itself. It can be indicated as the "crisis management" approach because it is not based on plans prepared in advance. Although the reactive approach still represents the most common response to drought events, there is an increasing awareness of its limitations since it implies last-minute decisions and leads to expensive actions, often with unsustainable environmental and social impacts.

The *proactive approach* consists of measures conceived and prepared according to a planning strategy rather than within an emergency framework. The proactive measures are devised and implemented before, during, and after the drought event. The measures are taken before the forecasting or initiation of a drought event to reduce the vulnerability to drought. They consist in *long-term actions* oriented to improve the reliability of the water supply system to meet future demands under drought conditions by a set of appropriate structural and institutional measures (Dziegielewski, 2003). The measures taken after the start of drought are *short-term actions*, which try to mitigate the impact of the particular drought event within the existing framework of infrastructures and management policies, on the basis of a contingency plan, studied in advance and adapted to the ongoing drought if necessary. Besides the category-based and approach-based classification of the drought mitigation measures, a correct analysis must take into account the affected societal sector, e.g., urban, agricultural, industrial, recreational, energy, or wildlife. In Table 9.1, a list of long-term and short-term measures is presented, subdivided into the three

Table 9.1 Classification of Measures for Coping with Drought

	Long-term measures	A.S.*	Short-term Measures	A.S.*
Supply increase	Augmentation of available resources through:		Use of additional water sources with low quality and/or high exploitation costs	U,A,I,R
	new surface reservoirs	U,A,I,R		
	interbasin and within-basin water transfers	U,A,I,R	Overexploitation of aquifers	U,A,I
	conveyance network for bidirectional exchanges	U,A,I	Increase diverted waters by relaxing ecological or recreational use	U,A,I
	reuse of treated wastewater	A,I	constraints (e.g., minimum instream	
	desalination of brackish or saline waters	U	flow, minimum lake level)	
	management of snow pack	A	Improvement of existing water systems efficiency	U,A,I
	control of evaporation losses	U,A,I	through leakage detection programs,	
	use of aquifers as ground water reserve	U,A,I	modified operation rules, etc.	
	rainfall augmentation	U,A,I		
Demand reduction	Dual distribution network for municipal use	U	Restriction of some municipal uses (car washing,	U
	Water recycle in industries	I	gardening)	
	Use of less water consumptive crops	A	Restriction of the irrigation of some crops (e.g., annual)	A
	Agronomic techniques for reducing water consumption	A	Pricing	U,A,I
			Public information campaign for voluntary water	U,A,I
	Sprinkler or drip irrigation	A	saving	
	Shift from irrigated to dry crops	A	Mandatory rationing	U,A,I
	Economic incentives for private investments in water conservation	U,A,I		

(*continued*)

Table 9.1 Classification of Measures for Coping with Drought (Continued)

	Long-term measures	A.S.*	Short-term measures	A.S.*
Impact minimiza-tion	Development of an early warning system	U,A,I	Temporary reallocation of water resources (on the	U,A,I
	Reallocation of water resources on the basis of water quality requirements	U,A,I	basis of assigned use priority)	
			Public aid to compensate loss of revenue	U,A,I
	Use of drought resistant crops	A	Tax relief (reduction or delay of payment	U,A,I
	Development of a drought contingency plan	U,A,I,R	deadline)	
			Rehabilitation programs	U,A,I
	Mitigation of economic and social impacts through voluntary insurance, pricing, and economic incentives	U,A,I		
	Education activities for improving preparedness to drought	U,A,I		

*Affected sector: U = urban; A = agricultural; I = industrial; R = recreation.
Source: Rossi, 2000.

categories of water supply increase, water demand reduction, and drought impact minimization. For each measure, the affected sectors are also indicated.

Despite the clear superiority of a proactive approach versus a reactive one, planning and implementing long- and short-term drought mitigation measures is difficult for several reasons, covering a large spectrum of scientific, institutional, and social factors. In particular, there is still an inadequate understanding of the natural drought phenomenon, and an inadequate development of appropriate tools aimed to assess the identified drought mitigation measures and to support, on a scientific basis, the decision-making process. Also, an early warning of water deficiency based on monitoring of hydrometeorological variables and water resources availability is often still lacking, probably due to the low appreciation of a proactive approach for coping with all natural hazards. Finally, it is difficult to quantify the impact of drought on different sectors (economy, environment, society), and evaluate the different perspectives of the stakeholders, since strong conflicts among different groups of interest often arise. From the institutional point of view, many legal and institutional constraints on the implementation of drought mitigation measures apply, while there is a lack of horizontal coordination among several water management agencies and a real vertical communication among different decision levels.

9.2.2 Multicriteria assessment of drought mitigation measures within the framework of a proactive approach

Implementation of a proactive approach to face drought problems for a given region or water supply system requires a procedure for selection of the best combination of long- and short-term measures. Application of MCA requires the definition of the alternatives to be considered and the assessment of the impacts of each alternative on the affected economic, social, and political sectors on the basis of the selected criteria. Therefore, the proposed procedure for the assessment of the drought mitigation alternatives, depicted in Figure 9.1, includes as a first step the assessment of the system vulnerability to drought in the current configuration and the computation of proper performance indices. Such an assessment can be carried out either with respect to a historical period or to generated hydrological scenarios, representing the future water supply availability. Then, the short- and long-term measures for coping with drought are identified among those exhibiting higher technical and economical feasibility, considering also political and institutional constraints. Also in this framework, a simulation model can be used to

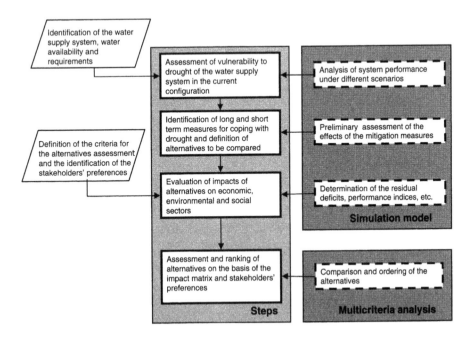

Figure 9.1 Proposed methodology for the identification and comparison of the drought mitigation measures.

determine the effects of the structural long-term measures on the water deficits reduction and to better determine the short-term measures to be adopted. In order to assess water supply systems performances, the simulation software SIMGES (Andreu et al., 1996) has been adopted.

The evaluation of the impacts of the alternatives is carried out by means of economic, environmental, and social criteria. The choice of the criteria depends on the nature of the vulnerability to drought of the present configuration of the water supply system as well as on the courses of action that water agencies, government at local and regional levels, and users have adopted during past droughts, or proposed as possible solutions to future drought threats.

The next step is the application of the multicriteria technique. A literature review has pointed out that most of the existing MCA techniques such as ELECTRE, Compromise Programming, Analytical Hierarchy Process, and PROMETHEE, have been applied to water supply system planning and management as well as to environmental decision problems (Raju and Pillai, 1999; Flug and Scott, 2000). In particular, drought mitigation measures have been analyzed through multicriteria techniques by Duckstein (1983) and by Munda et al. (1998).

Here the NAIADE model (Novel Approach to Imprecise Assessment and Decision Environments) developed by Munda et al. (1994) has been selected, as it is a discrete multicriteria method particularly oriented to evaluate alternatives for resources management and environmental protection. It allows either crisp, stochastic, or fuzzy measurements of the performance of an alternative with respect to a judgment criterion, thus it is very flexible for real world applications.

The main features of the NAIADE method can be synthesized as follows (Munda, 1995):

- The method is based on some aspects of the partial comparability axiom; in particular, a pairwise comparison between alternatives is carried out, which takes into account intensity of preference
- Use of a conflict analysis procedure allows a search for acceptable decisions, which earn a certain degree of consensus from the different interest groups

9.3 Application of the proposed procedure for the assessment of drought mitigation measures to the case studies

9.3.1 Sicilian case study (Italy): The Simeto system

9.3.1.1 Description of the system

The water system located in eastern Sicily around Catania Plain, shown in Figure 9.2, includes various agricultural, municipal, and industrial uses and is mainly supplied by a set of multipurpose plants for regulation and diversion of Salso-Simeto stream flows.

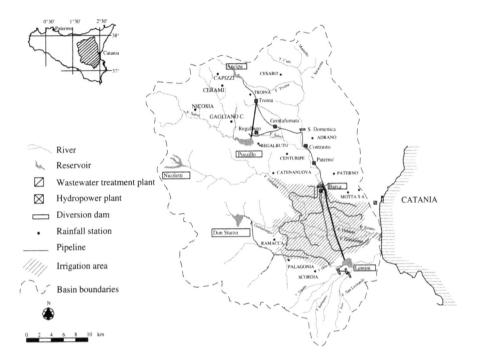

Figure 9.2 Simeto water supply system and Catania Plain irrigation district.

The system includes two reservoirs, Pozzillo on Salso river and Ancipa on Troina river (both tributaries of Simeto), three diversion dams located on the Simeto river (S. Domenica, Contrasto, and Ponte Barca), and six hydropower plants operated by the Electric Energy Agency (Enel). In addition, the Lentini reservoir is connected to the system via the Ponte Barca diversion dam on the Simeto river. The Ancipa reservoir has a design net capacity of $27.8 \cdot 10^6$ m^3, which is currently limited, due to structural problems, to $9.35 \cdot 10^6$ m^3. It regulates both the flows of the direct basin and of other tributaries, which are connected through a diversion canal. A small portion of the Ancipa releases are used to supply several municipalities in central Sicily, whereas the remaining portion is used for hydropower generation and irrigation purposes. The Pozzillo reservoir, which is mainly devoted to irrigation, has a current storage capacity of $123.0 \cdot 10^6$ m^3. Most of the releases are used for hydropower generation and irrigation of the main district of Catania Plain (irrigated area is about 18,000 h), whose water conveyance and distribution network is operated by the Land Reclamation Consortium 9 (LRC 9).

The Catania Plain irrigation district is particularly drought-prone, because of the significant variability of available water resources. For example, during the most severe dry events in past years, the releases dramatically decreased as shown in Table 9.2.

Since annual inflows to Pozzillo reservoirs have been below average for several years and consequently the storage capacity has not been fully used,

Table 9.2 Irrigated Areas and Releases for Catania Plain Irrigation District (Italy) during 1988–1991 and 1995 Droughts: A Comparison with the Average Values of the 1978-1987 decade

Year	Irrigated area (h)		Release		Irrigation duration (days)
	Citrus	Herbaceous crop	(10^6 m^3)	m^3/h	
Mean 1978–1987	14,340	2445	73	4350	139
1988	17,500	2750	80	3950	107
1989	17,000	—	23	1350	21
1990	16,500	—	6	364	7
1991	16,500	—	34	2060	41
1995	16,090	120	15	925	17

whereas annual inflows to Ancipa reservoir frequently exceed the reservoir storage capacity, a conduit has been recently built in order to transfer the winter spills from Ancipa to Pozzillo.

The Lentini reservoir has been built on the St. Leonardo river in order meet the demands of the irrigation districts managed by LRC 9 (Catania) and LRC 10 (Siracusa) and of the industrial areas of Siracusa and Catania. It partly regulates the flows of the Simeto river, through the Barca diversion, and the flows of four tributaries of St. Leonardo river: Dalle Cave, Trigona, Barbajanni, and Zena. It has been designed for a net storage capacity of $127 \cdot 10^6 \text{ m}^3$, although at present storage it is limited to $21.7 \cdot 10^6 \text{ m}^3$.

9.3.1.2   *Assessment of vulnerability to drought of the system in the current configuration*

The simulation of the system operation in the current configuration has been performed by SIMGES for the period 1959–1998 in order to acquire useful information about the system performances, as well as suggestions about the drought mitigation measures to be implemented. In the current system configuration, maximum capacity of Ancipa reservoir has been limited to $9.35 \cdot 10^6 \text{ m}^3$ due to the aforementioned structural problems, whereas Pozzillo maximum capacity is $123.0 \cdot 10^6 \text{ m}^3$. Lentini reservoir has been excluded from the simulation since its dam is still under testing. Furthermore, minimum storage volume constraints have been imposed for Ancipa, in order to safely meet the municipal demand.

Net irrigation requirements of Catania Plain irrigation district 9 have been estimated based on a study by the Land Reclamation Consortium of Catania Plain and from a study by CSEI Catania (1996). Net irrigation demand has been assumed equal to $85.0 \cdot 10^6 \text{ m}^3$/year, whereas the gross demand (including losses in the conveyance and distribution networks, consisting mostly of open channels) is $121.4 \cdot 10^6 \text{ m}^3$/year. Monthly municipal demands to Ancipa reservoir have been assumed constant and equal to $1.0 \cdot 10^6 \text{ m}^3$/month.

The results of the simulation are shown in Table 9.3 where several performance indices of the system in operation during the period 1959–1998 are

Table 9.3 Simeto Water Supply System: Performance Indices During 1959–1998

Municipal Use					Irrigation Use					
Releases		Temporal reliability	Releases		Temporal reliability	Average annual deficit	Max annual deficit	Sum of annual squared deficits	Deficit max duration	Deficit average duration
Mean	80%		Mean	80%						
(10^6 m^3)	(10^6 m^3)	(% years)	(10^6 m^3)	(10^6 m^3)	(% years)	(10^6 m^3)	(10^6 m^3)	$(10^6 \text{ m}^3)^2$	(years)	(years)
12.0	12.0	97.4	97.8	75.8	30.8	23.6	103.3	52165	10	3.9

reported. Since municipal use has priority over irrigation use, municipal demands are met almost every year, except for a very small deficit occurring in 1990. Consequently, mean annual release is practically equal to the demand, and volumetric and temporal reliability are very close to one. Highest irrigation annual deficit ($103.3 \cdot 10^6$ m³) occurs in 1990, during the most severe three-year historical drought. Temporal reliability for irrigation is equal to 30.8%, and the deficit average duration is 3.9 years, with a maximum of 10 years.

Both the duration and intensity of irrigation deficits suggest that long-term drought mitigation measures should be implemented in order to improve system capability to face severe drought conditions.

9.3.1.3 Alternatives for drought mitigation

Drought mitigation measures for the investigated system have been preliminarily identified among those exhibiting economical, political, and social feasibility. In addition, some long-term measures are already in the planning or execution stage. The considered long-term measures are oriented to increase the supply through water transfers and reuse of treated wastewater as well as to reduce irrigation demand through the replacement of existing distribution channel networks with pressure pipes in order to reduce losses, and through the development of small farm ponds. Short-term measures, on the other hand, are oriented to increase supply through overexploitation of ground water, to reduce demands by restricting the irrigation only to perennial crops, and to minimize the drought impacts by supporting the economy of the areas affected by drought through public aid.

More specifically, the selected long-term measures are:

L1: Water transfer from Ancipa reservoir to Pozzillo reservoir. Mean annual inflows to Ancipa exceed the capacity of the reservoir, causing significant winter spills. The measure consists in partly transferring such spills to Pozzillo reservoir, through a pipeline with maximum transport capacity of $8.0 \cdot 10^6$ m³/month.

L2: Modernization of the irrigation network. Currently, the conveyance and distribution networks for irrigation consist mostly of open channels, with low efficiency in terms of water losses. The measure consists of replacing such channels with pressure pipes, with an estimated loss reduction of $21.4 \cdot 10^6$ m³/year.

L3: Release for irrigation from Lentini reservoir. The relatively large capacity of Lentini has been designed to meet agricultural demands of the two irrigation districts LRC 9 (Catania Plain) and LRC 10 (Siracusa district) as well as the industrial demand of Siracusa and Catania industrial areas. Although the delays in the construction of the reservoir and of the related conveyance pipelines have limited the full utilization of the facility, winter waters are already diverted by the Simeto at Barca diversion dam and transferred to Lentini. The measure consists of supplying the irrigation area of LRC 9 from Lentini up to a maximum volume of $21.5 \cdot 10^6$ m³/year, the irrigation area of Siracusa district

(LRC 10) for $18.5 \cdot 10^6$ m³/year and the industrial areas of Siracusa and Catania for $10.0 \cdot 10^6$ m³/year and $20.0 \cdot 10^6$ m³/year respectively.

L4: Treated wastewater reuse from Catania plant. Although rejected in the past, the possibility to use marginal waters for irrigation is now becoming more and more appealing. The measure consists of constructing the facilities necessary to use wastewater from the Catania treatment plant for the irrigation of a portion of the Catania Plain, up to a volume of $8.4 \cdot 10^6$ m³.

L5: Construction of small reservoirs by farmers. The use of small private reservoirs by farmers is currently a common practice, due also to economic incentives in the past for their development. The construction of new small reservoirs could provide an additional storage capacity of $8.0 \cdot 10^6$ m³, which farmers could use as a strategic reserve for drought periods.

Figure 9.3 shows the scheme of the water supply system, where the considered long-term measures are highlighted.

The following short-term measures have been identified, based on an analysis of management of past droughts:

S1: Supplementary resources from ground water and ponds. Ground water withdrawals and storage in private ponds is a common practice

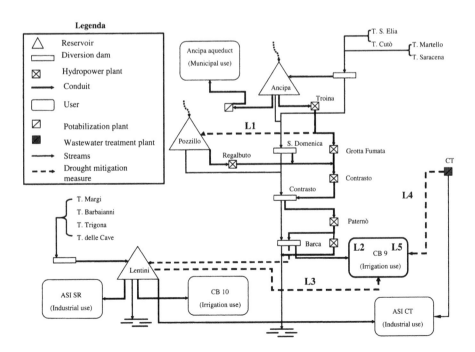

Figure 9.3 Simeto water supply system: Scheme with long-term measures.

adopted by farmers to compensate for the deficit in supply from collective irrigation networks. Although there is a lack of detailed information about the consistency of such phenomenon, it can be assumed that an additional water supply of $20.0 \cdot 10^6$ m^3 can be made available during droughts.

S2: Management criteria to face water scarcity. When severe droughts occur, it is essential to ensure irrigation for perennial crops in order to avoid damages to the trees and consequent capital losses. The measure consists of restricting irrigation to perennial crops, excluding irrigation of annual crops, increasing vigilance to prevent water thefts, and ensuring the respect of irrigation turns.

S3: Natural calamity aids. During droughts that occurred in the past, regional and national governments have already taken actions to reduce the economic effects of drought recognized as natural calamities, such as public refund for damages, low-interest rate loans, and tax relief. Similar measures are expected for future drought events.

The alternatives for drought mitigation in the Catania Plain, combination of long- and short-term measures, are summarized in Table 9.4, where the long-term measure L0 refers to the system in the current configuration. It

Table 9.4. Simeto Water Supply System: Selected Alternatives for Drought Mitigation

		Alternatives							
		A	B	C	D	E	F	G	H
	Long-term measures								
L0	System in the current configuration	X							
L1	Water transfer from Ancipa res. to Pozzillo res.		X					X	X
L2	Modernization of the irrigation network			X					
L3	Release for irrigation from Lentini reservoir				X			X	
L4	Treated wastewater reuse from Catania plant					X			X
L5	Construction of small reservoirs by farmers						X		
	Short-term measures								
S1	Supplementary resources from gound water and ponds	X	X	X	X	X	X	X	X
S2	Management criteria to face water scarcity	X	X	X	X	X	X	X	X
S3	Natural calamity aids	X	X	X	X	X	X	X	X

can be inferred from the table that the alternatives consist of all of the three short-term measures and of one or more long-term measures.

9.3.1.4 Performance of the system under different drought mitigation measures

Assessment of the performance of the system considering the different measures for coping with drought has been carried out in terms of releases and deficits, by considering as alternatives the combinations of one or more long-term measures plus the short-term measure S1. Short-term measures S2 and S3, on the other hand, have not been considered explicitly in the simulation since they do not affect the releases and deficits, but are rather oriented to reduce the consequent economic losses. Their effects, however, will be taken into account within the multicriteria assessment of the alternatives.

As an example, in Figure 9.4 the annual irrigation releases to LRC 9 consequent to the long-term measures L1 + L4 (alternative H) respectively are depicted, along with the additional releases due to short-term measure S1 and the residual deficits.

After simulation of each alternative, performance indices of the behavior of the system have been computed. In particular, reliability indices (both temporal and volumetric), resilience index (computed as the inverse of the average deficit period duration), and various vulnerability indices have been used here. In addition, the sustainability index proposed by Loucks and Gladwell (1999) has also been computed. The indices related to the irrigation demands LRC 9 and LRC 10 are reported in Table 9.5 for each considered alternative. It can be inferred that the system in the status quo performs poorly in terms of all the indices, whereas all the other options significantly increase temporal reliability and sustainability, computed as a function of the temporal reliability itself, the resiliency, the maximum annual deficit, and the maximum deficit duration. Furthermore, all alternatives reduce the sum of the squared annual irrigation deficits, besides the number of years where the annual deficit exceeds 25% of the annual demand.

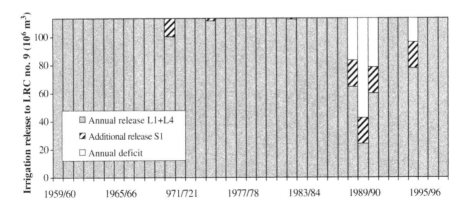

Figure 9.4 Alternative H: Irrigation releases to Catania Plain.

Table 9.5 Catania Plain Irrigation District: Performance Indices for the Different System Configurations

Simulation	Annual demand		Temporal reliability at	Sum of squared annual deficits	Irrigation use No. of years with deficit > 25% demand	Resilience Pes.	Vulnerability Max annual def. MAD	Max def. duration MDD	Sustainability index S(*)
		(10^6 m^3)	(% years)	$(10^6 \text{ m}^3)^2$	(−)	(years^{-1})	(10^6 m^3)	(years)	(−)
L0 + S1	LRC 9	121.4	0.590	24682	6	0.63	83.3	3	0.108
L1 + S1	LRC 9	121.4	0.821	14511	4	0.71	83.3	3	0.169
L2 + S1	LRC 9	100.0	0.949	4660	1	0.50	63.5	2	0.164
L3 + S1	LRC 9	121.4	0.949	4777	1	0.50	63.7	2	0.214
	LRC 10	18.5	1	0	0	—	0	0	—
L4 + S1	LRC 9	113.0	0.897	8667	3	0.50	75.7	3	0.137
L5 + S1	LRC 9	113.4	0.897	8777	3	0.50	75.9	3	0.137
L1 + L3 + S1	LRC 9	113.0	0.949	4777	1	0.50	63.7	2	0.214
	LRC 10	18.5	1	0	0	—	0	0	—
L1 + L4 + S1	LRC 9	113.0	0.923	8489	3	0.50	75.9	2	0.144

$$^{(*)} S = At \cdot Res \cdot \left(1 - \frac{MAD}{Annual\ Demand}\right)\left(1 - \frac{MDD}{No.\ of\ years}\right)$$

Municipal and industrial use demands are generally satisfied, since a greater priority has been given to these uses. Consequently, the corresponding performance indices (here not reported for the sake of brevity) reflect an almost 100% reliability, a very high resilience and very low vulnerability in all cases.

9.3.1.5 Identification of evaluation criteria and stakeholders

The criteria to assess each alternative have been chosen in order to take into account properly the different economic, environmental, and social consequences of drought measures adopted in each alternative. Accordingly, three types of criteria have been selected:

Economic criteria.

1a. Construction costs: These include the capital costs necessary for the implementation of the long-term structural measures. Such costs are computed on the basis of the design project of the structure if existing, otherwise indirectly on the basis of standard unit costs.

1b. Operation and maintenance costs: These costs refer to the operation and maintenance costs of the system infrastructures, as well as to recovery costs. They are considered as the present worth of annual fixed percentages of the capital costs of structures and electromechanical parts.

1c. Short-term measures costs: These costs refer to the present worth of the expected cost of short-term measures. They are determined as the cost of the additional water from ground water plus the cost of the increased vigilance of the conveyance and distribution network and of the public refunds.

1d. Damages to perennial crops: Damages may occur to perennial crops in case of severe water deficits. As a proxy for the estimation of such damages, the percentage of years when the residual annual deficits are greater than the 25% of the demand is considered here.

Environmental criteria.

2a and 2b. Failures to meet minimum storage volumes: Minimum storage volumes for environmental purpose are considered for Pozzillo $(8.0 \cdot 10^6 \text{ m}^3)$ and Ancipa $(1.5 \cdot 10^6 \text{ m}^3)$ reservoirs. These criteria consider the percentage of the months when such minimum storage volumes are not met in Pozzillo (criteria 2a) and Ancipa (criteria 2b).

2c. Sustainability: This criterion takes into account the different degrees of each alternative's sustainability. In particular, the alternatives that do not involve ground water overexploitation but favor the use of renewable sources.

2d. Reversibility: This takes into account the reversibility of the actions, i.e., the possibility to restore the initial conditions of the system with respect to the economic or environmental feasibility.

Social criteria.

3a. System vulnerability: Since higher concentrated deficits will cause more severe effects on society, a water system can be considered less vulnerable to drought if it tends to distribute deficits over time. Here the system vulnerability is computed as the sum of the squared annual deficits.

3b. Temporal reliability: This criterion considers the percentage of years when a given demand is fully satisfied.

3c. Realization time of the infrastructures: This takes into account the time of realization of the infrastructures, distinguishing alternatives whose realization is not feasible in a short period and alternatives with fast implementation.

3d. Employment increase: This criterion takes into account the consequence of an alternative with the respect of the employment increment in the construction, operation, and maintenance period.

In Table 9.6, the evaluation criteria grouped for affected sectors and related units are reported.

Among the several groups of stakeholders that are affected by or play an active role in the management of the system, the following have been identified: G1 — Irrigation Management Agency (Land Reclamation Consortia 9

Table 9.6 Simeto Water Supply System: Assessing Criteria and Units

Economic Criteria	Units
1a Construction costs of infrastructures	Millions of €
1b Operation and maintenance cost of infrastructures	Millions of €
1c Short-term measures cost	Millions of €
1d Damages to perennial crops	Number of years with deficits greater than 25% of demand

Environmental Criteria	Units
2a Failure to meet minimum storage in Pozzillo	% months
2b Failure to meet minimum storage in Ancipa	% months
2c Sustainability of the measure	Qualitative
2d Reversibility of the measure	Qualitative

Social Criteria	Units
3a Vulnerability of the system	Sum of squared deficits
3b Temporal reliability	% of years
3c Realization time	Qualitative
3d Employment increase	Qualitative

and 10), G2 — Farmers of Catania Plain district, G3 — Hydroelectric Power Agency (ENEL), G4 — Industries, G5 — Environmentalists.

9.3.1.6 Impact analysis

As previously mentioned, impacts of each alternative have been assessed by simulating the system considering the long-term measures and by reducing the consequent annual deficits to take into account the increase of supply given by short-term measure S1, up to a volume of $20.0 \cdot 10^6$ m^3/year. When residual deficits exceed $15.0 \cdot 10^6$ m^3/year short-term measures S2 (mandatory limitation of irrigation to perennial crops) and S3 (public aid to drought affected stakeholders) have been introduced in the analysis.

In evaluating the impacts, the following assumptions have been made:

- Only one third of the capital cost 1a and the operation and maintenance costs 1b of alternative L2 have been considered, to take into account the fact that modernization of the whole conveyance and distribution network is necessary, regardless of drought mitigation, in order to ensure good performance of the irrigation system even in normal (nondrought) periods.
- In the computation of the temporal reliability, only irrigation use (LRC 9) has been taken into account, since both the municipal demand and the industrial demands (Catania and Siracusa industrial areas) are always fully met.

In Table 9.7 the impact analysis matrix is reported.

The resulting ranking of the alternatives obtained through NAIADE application is depicted in Figure 9.5. From the figure it can be inferred that the alternatives with the highest ranking are D (release for irrigation from Lentini reservoir as long-term measure, besides the short-term measures S1 + S2 + S3), G (water transfer from Ancipa to Pozzillo reservoir plus release for irrigation from Lentini reservoir), and C (modernization of the irrigation network). The current system configuration (alternative A) along with the construction of small reservoirs (alternative F) have the lowest ranking.

9.3.1.7 Conflict analysis

Alternatives D, G, and C perform on the whole better than the others, and therefore in principle they should be preferred. However, ranking of the alternatives represents only partial information, since it is known that in a political decision-making process the final solution is often a result of a compromise between the different stakeholders.

From the preference matrix, which expresses the preferences of each stakeholder with respect to each alternative, it is possible to build the dendrogram of coalitions shown in Figure 9.6, which describes the process of coalition formation and the related agreement levels.

Table 9.7 Simeto Water Supply System: Impact Analysis Matrix

							Criteria					
Alternatives	1a (10⁶ Euro)	1b (10⁶ Euro)	1c (10⁶ Euro)	1d (years)	2a (%months)	2b (% months)	2c Qual	2d Qual	3a (10⁶ m³)²	3b (% years)	3c Qual	3d Qual
A	0	0	142.5	6	4.1	29.7	EB	P	24,682	59.0	P	VB
B	10,119	7,837	55.9	4	13.3	16.2	B	VG	14,511	82.1	VG	MLG
C	81,978	73,616	22.4	1	13.2	5.6	G	G	4660	94.9	VB	P
D	12,867	12,397	23.9	1	17.1	6.0	B	M	4777	94.9	M	G
E	7,033	6,776	47.9	3	13.5	11.8	P	G	8667	89.7	MLG	M
F	4,132	5,596	48.2	3	13.5	12.0	M	VB	8777	89.7	G	MLB
G	22,986	20,234	23.9	1	17.1	6.0	MLB	MLB	4777	94.9	M	VG
H	17,152	14,613	44.9	3	10.7	13.5	G	G	8489	89.7	MLG	G

Note: EB = extremely bad, VB = very bad, B = bad, MLB = more or less bad, M = moderate, MLG = more or less good, G = good, VG = very good, P = perfect.

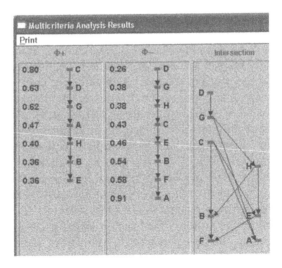

Figure 9.5 Simeto water supply system: Ranking of alternatives.

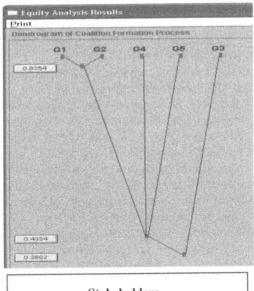

Stakeholders
G1: Irrigation Management Agency (LRC)
G2: Farmers of Catania Plain District
G3: Hydroelectric Power Agency
G4: Industries
G5: Environmentalists

Figure 9.6 Simeto water supply system: Process of coalition formation and related agreement levels.

The coalition formation process points out that the first coalition is formed by the groups G1 and G2, namely the Irrigation Management Agency and the farmers of the Catania Plain district, which is quite obvious, since they have basically the same goals. This coalition only vetoes alternatives E, F, and A, namely the reuse of treated wastewater, the construction of small reservoirs by farmers, and status quo. Therefore, the alternatives preferred from the point of view of this group are still D, G, and C.

However, from previous experience, it could be argued that to reach a solution, a wider consensus of interest groups is required. Considering the coalition of groups G1, G2, G4, and G5, the vetoed alternatives would be D, G, E, F, and A, which narrows the set of eligible alternatives to C, H (water transfer from Ancipa to Pozzillo reservoir plus treated wastewater reuse), and B (water transfer from Ancipa to Pozzillo reservoir). Since alternative B performed poorly in the ranking, the feasible set is further reduced to C and H. However, since in the first phase of the procedure alternative C ranked better than alternative H, the modernization of the irrigation network seems to be eventually the most preferable alternative.

9.3.2 Sardinian case study (Italy): The Flumendosa–Campidano system

9.3.2.1 Description of the system

The application to the Sardinian case study has been carried out by CINSA (Interdepartmental Center for Environmental Science and Technology), University of Cagliari. The Flumendosa–Campidano water system shown in Figure 9.7 supplies most of the southern part of Sardinia. It includes the basins of several rivers that flow to the eastern part of the gulf of Cagliari. The system has climatic features typical of the Mediterranean regions, with strongly variable precipitations from one year to another, and presents a high level of complexity due also to the interconnections with other systems. It is a multipurpose system (municipal, industrial, and irrigation uses) operated by the Flumendosa Water Authority.

The main water supply source of the system is represented by three reservoirs connected by gravity galleries: Flumineddu, Flumendosa, and Mulargia, with a total storage capacity of about $670 \cdot 10^6$ m^3. The aquifers contribution to the system can be considered negligible.

A multicriteria analysis has been performed to evaluate the consequences on drought mitigation of the works considered in phase 1 of the Regional Basin Plan, with particular reference to the effects produced by their progressive developments.

9.3.2.2 Definition of long-term measures and alternatives

The drought mitigation measures are distinguished, on the basis of the time horizon, in long-term measures (L_i) and short-term measures (S_i) as follows:

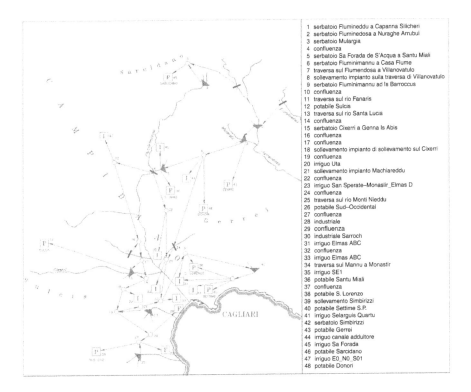

1 serbatoio Flumineddu a Capanna Silicheri
2 serbatoio Fluminedosa a Nuraghe Arrubui
3 serbatoio Mulargia
4 confluenza
5 serbatoio Sa Forada de S'Acqua a Santu Miali
6 serbatoio Fluminimannu a Casa Flume
7 traversa sul Flumendosa a Villanovatulo
8 sollevamento impianto sulla traversa di Villanovatulo
9 serbatoio Fluminimannu ad Is Barroccus
10 confluenza
11 traversa sul rio Fanaris
12 potabile Sulcis
13 traversa sul rio Santa Lucia
14 confluenza
15 serbatoio Cixerri a Genna Is Abis
16 confluenza
17 confluenza
18 sollevamento impianto di sollevamento sul Cixerri
19 confluenza
20 irriguo Uta
21 sollevamento impianto Machiareddu
22 confluenza
23 irriguo San Sperate–Monaslir_Elmas D
24 confluenza
25 traversa sul rio Monti Nieddu
26 potabile Sud–Occidental
27 confluenza
28 industriale
29 confluenza
30 industriale Sarroch
31 irriguo Elmas ABC
32 confluenza
33 irriguo Elmas ABC
34 traversa sul Mannu a Monastir
35 irriguo SE1
36 potabile Santu Miali
37 confluenza
38 potabile S. Lorenzo
39 sollevamento Simbirizzi
40 potabile Settime S.P.
41 irriguo Selarguis Quartu
42 serbatoio Simbirizzi
43 potabile Gerrei
44 irriguo canale adduttore
45 irriguo Sa Forada
46 potabile Sarcidano
47 irriguo E0_N0_S01
48 potabile Donori

Figure 9.7 The Flumendosa–Campidano water supply system.

Long-term measures.

L1: Integration of the existing diversion dam in the southern area of the system with the Monti Nieddu dam at Sa Stria ($35 \cdot 10^6$ m³) and the diversion dam of Is Canargius upstream from the main dam.

L2: Realization of the reservoir of Monte Perdosu ($78.91 \cdot 10^6$ m³), taking the place of the diversion dam of S'Isca Rena on Rio di Quirra (in the central-eastern area of the system). The new reservoir will be connected in a bidirectional form with the Mulargia reservoir, and with the new diversion dam on Rio Quirra. Furthermore, transfers toward the local civil and irrigation demands of Muravera are also foreseen.

L3: Construction of a reservoir on Rio Foddeddu ($42.4 \cdot 10^6$ m³) in the eastern area of the system, planned to serve all the municipal demands of the aqueduct systems 26-27-28 (as indicated in the Regional Water Plan), the irrigation demands of Cea-Tennori and Tortoli and the industrial complex of Arbatax. The system is further integrated by the Pramaera diversion dam, connected to the S. Lucia reservoir. The developing of the Arbatax industrial area is also included in the measure, with the consequent increase of water demand of $12 \cdot 10^6$ m³/year.

L4: Integration of the management of the Flumendosa-Campidano system with the Cixerri system, through the bidirectional connection with the Cixerri reservoir. Besides, the diversion dams of S'Acqua Fisca, Rio Figu, and San Marco will be connected to the Cixerri–Flumendosa–Campidano system.

L5: Realization of a desalination plant in the city of Cagliari, whose capacity is estimated in $30 \cdot 10^6$ m³/year. Moreover, the reuse of $9.37 \cdot 10^6$ m³/year obtained by the wastewater treatment plant of Monastir, Serranmana, and the CASIC industrial area is foreseen. Considering this new facility and the contribution of the Is Arenas plant, the total water reuse will reach $31.15 \cdot 10^6$ m³/year.

Short-term measures.

S1: Exploitation of the ground water resources for the partial fulfillment of the municipal demand. In the Regional Basin Plan such contribution is estimated in $21.83 \cdot 10^6$ m³/year.

S2: Reduction of the ENEL (Electricity Agency) water assignment in High Flumendosa hydroelectric plant, with the consequent elimination of the storage volumes constraints for the reservoir of Sicca d'Erba.

S3: Water transfer from the wells of Campo Pisano to the multisector system of Flumendosa–Campidano.

S4: Integration of the management of the Flumendosa–Campidano system with the eastern regional scheme.

S5: Water transfer from the Tirso system through a bidirectional connection between the diversion dam on Rio Mogoro, in the Tirso basin, and the Sa Forada reservoir in the middle Flumendosa–Campidano area.

The considered alternatives (defined in agreement with the preliminary version of the Regional Basin Plan) are shown in Table 9.8, where L0 indicates the water supply system in the current configuration.

9.3.2.3 Identification of evaluation criteria and stakeholders
The following economic, environmental, and social criteria have been taken into account:

Economic criteria.

1a: Investments costs for construction of infrastructures referred only to the new works needed to implement the alternatives.

1b: Water demands for municipal use (1b.1), industrial use (1b.2), and irrigation use in the high (1b.3) and middle-lower (1b.4) Flumendosa–Campidano system.

1c: Increase of demand for municipal use (1c.1), industrial use (1c.2), and irrigation use in the high (1c.3) and middle-lower (1c.4) system.

1d: Reduction of the irrigation demand in the high (1d.1) and middle-lower (1d.2) system.

Table 9.8 Flumendosa-Campidano Water Supply System: Selected Alternatives for Drought Mitigation

		Alternatives							
		A	B	C	D	E	F	G	H
	Long-term measures								
L0	System in the current configuration	X	X	X	X	X	X	X	X
L1	Connection of two intakes to the system				X	X	X	X	X
L2	Realization of a new reservoir for municipal and irrigation supply					X	X	X	X
L3	Inclusion of two reservoirs for municipal and industrial supply						X	X	X
L4	Connection with the Cixerri system through one reservoir and several intakes							X	X
L5	Realization of a desalination plant and wastewaters reuse								X
	Short-term measures								
S1	Groundwater exploitation	X	X	X	X	X	X	X	X
S2	Elimination of minimum storage constraints in Sicca d'Erba reservoir	X	X	X	X	X	X	X	X
S3	Use of Campo Pisano wells	X	X	X	X	X	X	X	X
S4	Integration of the system management with the Eastern regional scheme		X	X	X	X	X	X	X
S5	Water transfer from the Tirso system			X	X	X	X	X	X

1e: Irrigation reduction of multiannual crops (triggered when the cut in the irrigation release is over 25% of the demand) in the high (1e.1) and middle-lower (1e.2) system.

Environmental criteria.

2a: Violation to instream flow requirements from the reservoirs.
2b: Environmental impacts of the different actions (estimated qualitatively as a function of the type of intervention).

Social criteria.

3a: Volumetric vulnerability of the system, computed as the difference between the irrigation demands and the respective net releases in the high (3a.1) and middle-lower (3a.2) system.
3b: Realization time of the infrastructures.
3c: Reversibility of the interventions.

The evaluation criteria grouped for affected sectors are reported in Table 9.9.

Table 9.9 Flumendosa–Campidano Water Supply System:
Assessing Criteria and Units

	Economic Criteria	**Units**
1a	Construction costs of infrastructures	Millions of €
1b	Water demand for municipal use (1b.1), industrial use (1b.2), irrigation use in the high (1b.3) and middle-lower (1b.4) system	10^6 m³/year
1c	Increase of demand for municipal use (1c.1), industrial use (1c.2), irrigation use in the high (1c.3) and middle-lower (1c.4) system	10^6 m³/year
1d	Coefficient of reduction of the irrigation demand in the high (1d.1) and middle-lower system (1d.2)	%
1e	Area of nonirrigated multiannul crops in the high (1e.1) and middle-lower system (1e.2)	Hectares
	Environmental Criteria	**Units**
2a	Violation to instream flow requirements from reservoirs	10^6 m³/ month
2b	Environmental impact of different types of actions	Qualitative
	Social criteria	**Units**
3a	Difference between the irrigation demands and net releases in the high (3a.1) and middle-low system (3a.2)	10^6 m³/year
3b	Realization time of the largest infrastructures	years
3c	Reversibility of the intervention	Qualitative

9.3.2.4 *Assessment of alternatives (impact and conflict analysis)*

After simulation of the system in the different configurations and management alternatives, the impact matrix reported in Table 9.10 has been obtained. A sensitivity analysis has been carried out by varying the parameter α of NAIADE, which is basically a threshold parameter of the fuzzy membership functions (NAIADE Manual, 1996). The selected ranking for different values is depicted in Figure 9.8.

The following dominance relationships hold: H > G > F > E (group of the best alternatives), C > A and B > A (group of the worst alternatives), while the position of the alternative D may change.

The increase of the preference threshold reduces the dominance of the alternatives F and G on D. The results obtained show that the group of the alternatives F, G, H gives a better performance in terms of efficiency. Concerning the dominance inside this group, the alternative H dominates the alternatives

Table 9.10 Flumendosa–Campidano Water Supply System: Impact Analysis Matrix

Criteria	Alternatives							
	A	B	C	D	E	F	G	H
1a (10⁶ Euro)	0.0	0.0	62.0	25.8	129.1	49.1	26.3	82.7
1b.1 (10⁶ m³/yr)	100.07	102.62	102.62	102.62	103.72	103.72	103.72	103.72
1b.2 (10⁶ m³/yr)	21.05	21.05	21.05	21.05	21.05	33.05	33.05	33.05
1b.3 (10⁶ m³/yr)	25.03	25.03	25.03	25.03	25.03	25.03	25.03	25.03
1b.4 (10⁶ m³/yr)	235.84	235.84	235.84	247.52	259.2	259.2	259.2	269.95
1c.1 (10⁶ m³/yr)	0.0	2.55	2.55	2.55	3.65	3.65	3.65	3.65
1c.2 (10⁶ m³/yr)	0.0	0.0	0.0	0.0	0.0	12	12	12
1c.3 (10⁶ m³/yr)	0.0	0.0	0.0	0.0	5.0	5.0	5	5
1c.4 (10⁶ m³/yr)	0.0	0.0	0.0	11.68	23.36	23.36	23.36	34.11
1d.1 (%)	83	86	86	86	86	85	84	90
1d.2 (%)	58	63	78	78	91	93	88	95
1e.1 (Ha)	70	0	0	0	0	0	0	0
1e.2 (Ha)	22311	19430	0	0	0	0	4113	0
2a (10⁶ m³/month)	2.06	2.17	2.18	2.22	3.39	3.34	3.21	2.76
2b (Qualitative)	G	G	MLG	B	B	B	MLG	B
3a.1 (10⁶ m³/yr)	4.25	3.51	3.51	3.51	3.51	3.73	3.95	2.5
3a.2 (10⁶ m³/yr)	99.06	86.47	52.03	53.35	23.82	17.91	30.37	13.35
3b (Years)	—	0	1	3	11	11	1	3
3c (Qualitative)	B	G	M	B	B	B	G	B

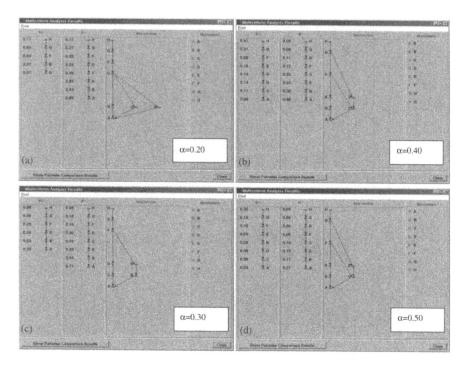

Figure 9.8 Flumendosa–Campidano water supply system: Ranking of alternatives for different values of α.

F and G. The alternatives A, B, C rank the lowest positions in the global order with the alternatives B and C dominating the alternative A. The results obtained confirm that the efficiency in terms of drought mitigation is maximum when the Flumendosa–Campidano system is fully integrated in a single regional scheme.

Due to different priorities assigned to each use, a conflict arises between the different stakeholders. Higher priority has been given to municipal, irrigation, and industrial use respectively. Regarding the irrigation demand, it is necessary to distinguish the interests of the middle-lower area as opposed to the high area. The following interest groups have been identified: G1 — Municipal users, G2 — Industrial users, G3 — Farmers of the high system, and G4 — Farmers of the middle-lower system.

The conflict analysis has been carried out assigning the preferences of the involved interest groups on the basis of the simulations results. The results of the conflict analysis are shown in Figure 9.9.

For a level of compromise equal to 0.6983, the coalition between the groups G1 and G2 puts a veto on the alternatives A, C, D. The alternatives B and F are instead vetoed for the same level of compromise respectively by group G4 and G3. For a level of compromise equal to 0.6030, the agreement is reached between groups G4, G2, and G1. If an agreement between all the stakeholders is desired, the level of compromise to be chosen is 0.5774. However, the levels of compromise reached are not significantly different.

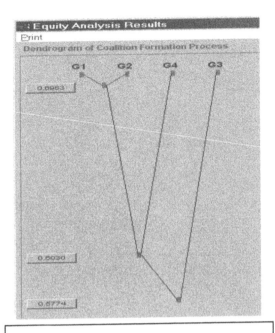

<table>
<tr><td>Stakeholders</td></tr>
</table>

G1: Municipal users
G2: Industrial users
G3: Farmers of the high system
G4: Farmers of the middle-lower system

Figure 9.9 Flumendosa–Campidano water supply system: Process of coalition formation and related agreement levels.

This shows that there is a substantial agreement between the interest groups with respect to the alternatives proposed.

9.3.3 Spanish case study: The Júcar system

9.3.3.1 Description of the Júcar system

The application of the proposed procedure for multicriteria assessment to the Spanish case study has been carried out by the Universidad Politécnica de Valencia, Department of Civil and Environmental Engineering. The Júcar basin (Figure 9.10), with a surface of 22.378 km², is located in the central-eastern sector of Spain, is characterized by extremely dry summer periods as opposed to wet fall-spring periods. The water supply system includes two main rivers: Júcar and Cabriel (its main tributary), and three big regulation reservoirs: Alarcón, Contreras (in the upper basin), and Tous (middle-low basin). The average flow in the period 1940/41–2000/01 has been $1340 \cdot 10^6 \, m^3/year$ from the basin down to Tous reservoir.

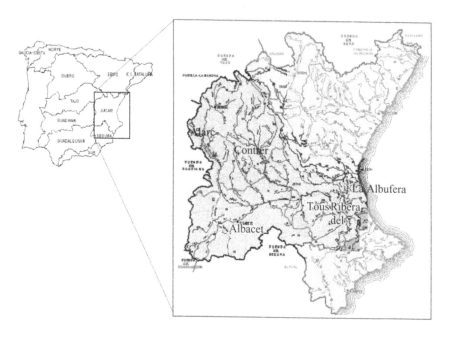

Figure 9.10 Location of the Júcar basin in the Spanish territory.

Uses for irrigation include the Ribera del Júcar irrigation area, located at the coastal zone, as well as the new intensive demand of Albacete at the middle sector of the basin, which is supplied by ground water extraction from the Mancha Oriental Aquifer. There are also domestic demands supplied by the resources of the Júcar basin, to the towns of Valencia, Albacete, and Sagunto, and also some water transfers to other deficient systems outside the basin, like the transfer to the Marina Baja tourist area.

9.3.3.2 Definition of mitigation measures and alternatives

As a first step of the analysis, the short- and long-term measures for coping with droughts have been identified. The selected short-term measures include the use of drought wells (i.e., wells that are operated only in deficit situations) located in the area of Ribera Alta del Júcar during the summer or the winter (measures S1 and S3 respectively) and the application of different levels of demand restrictions in the areas irrigated with surface water or both surface and ground water (measure S2).

Long-term measures have been identified among those currently under execution or in a preproject phase. They include the reuse for irrigation of wastewater from the urban settings in the area of Ribera del Júcar (measure L1), the modernization of the irrigation network, i.e., of the main ditch Acequia Real del Júcar and the whole distribution network that supplies the fields of Ribera Alta (measure L2), the development of a desalination plant in the Marina Baja area (measure L3), and the electrification of drought wells at Ribera Alta del Júcar, previously operated with diesel motors (measure L4).

Table 9.11 Júcar Water Supply System: Selected Alternatives for Drought Mitigation

		Alternatives							
		A	B	C	D	E	F	G	H
	Long-term measures								
L0	System in the current configuration	X	X	X	X				
L1	Reuse of wastewater for irrigation Ribera del Júcar					X			
L2	Modernization of the irrigation network						X		
L3	Desalination plant in the Marina Baja area							X	
L4	Electrification of drought wells								X
	Short-term measures								
S1	Use of drought wells (Ribera Altadel Júcar)	X			X	X	X	X	X
S2	Application of restrictions on irrigation			X		X	X	X	X	X
S3	Increase of use of ground water				X	X	X		X	X

Then the selected measures have been grouped into management alternatives, as reported in Table 9.11, where L0 indicates the water system in the current configuration.

9.3.3.3 Identification of evaluation criteria and stakeholders

Economic, environmental, and social criteria have been adopted for the analysis (Table 9.12). In detail, these criteria include:

Economic criteria.

1a: Construction costs, related to new infrastructures.
1b: Operation and maintenance costs on annual basis.
1c: Short-term costs, related to the occasional use of the drought wells.
1d: Damages due to restrictions, evaluated in an indirect way, as the costs to obtain water from other sources.

Environmental criteria.

2a: Failures of the environmental flows at the Albufera wetland, indicating how many times the irrigation returns to the wetland of La Albufera are less than $3 \ m^3 \cdot 10^6/month$ ($36 \ m^3 \cdot 10^6/year$).
2b: Failures of the environmental flows at the Júcar middle sector, indicating how many times the flows circulating through the middle sector of Júcar river are less than $1 \ m^3/s$.

Table 9.12 Júcar Water Supply System: Evaluation Criteria and Measurement Units

Economic Criteria	Units
1a Construction costs of infrastructures	Millions of Euro dollars
1b Operation and maintenance costs	Millions of Euro dollars
1c Short-term costs	Millions of Euro dollars
1d Damages due to water restrictions	Millions of Euro dollars

Environmental Criteria	Units
2a Failures of the environmental flows at the Albufera wetland	Number of failures
2b Failures of the environmental flows at the Júcar middle sector	Number of failures
2c Impact sustainability (River losses towards aquifer)	$10^6\,m^3$/year

Social Criteria	Units
3a System vulnerability 1 year	% of annual demand
3b System vulnerability 2 year	% of annual demand
3c System vulnerability 10 year	% of annual demand

2c: Impact sustainability (river losses towards aquifer), representing the reduction of the natural discharge from the aquifer Mancha Oriental to Júcar river due to groundwater extraction from the aquifer.

Social criteria.

3a: System vulnerability at one (3a.1), two (3a.2), and 10 (3a.3) years, expressed as the percentages over the annual demand of the maximum deficit in one, two, or 10 consecutive years, obtained as the average of deficits at the areas of Ribera Alta and Baja.

A large number of stakeholders have an active role in the water management of the Júcar system, and therefore only a selection of the groups directly affected by the management alternatives analyzed in this study have been included in the coalition formation analysis. The considered stakeholders include: G1 — Tourist water use board of the Marina Baja area (Consorcio de Abastecimiento y Saneamiento de Aguas de la Marina Baja), that benefits from the restrictions of surface supply (substituted with ground water) to the irrigation area of Ribera Baja; G2 — Farmers of Acequia Real del Júcar (Ribera Alta); G3 — Farmers of Ribera Baja del Júcar; G4 — Farmers of Canal Júcar-Turia; G5 — Iberdrola Hydroelectric Company; G6 — Environmental organizations and public opinion; and G7 — Domestic supply users of Valencia.

9.3.3.4 Assessment of alternatives (impact and conflict analysis)

After simulation of the system in the different configurations and management alternatives, the impact matrix reported in Table 9.13 was obtained.

Table 9.13 Júcar Water Supply System: Impact Analysis Matrix

					Criteria					
Alternatives	1a (10⁶ Euro)	1b (10⁶ Euro)	1c (10⁶ Euro)	1d (10⁶ Euro)	2a (n. of failures)	2b (n. of failures)	2c (10⁶ m³/year)	3a (% annual demand)	3b (% annual demand)	3c (% annual demand)
A	0	1.8	341.9	287.1	75	133	287.6	64.0	115.9	274.7
B	0	—	348.3	335.4	48	110	285.7	67.8	124.9	330.7
C	0	1.2	348.7	300.3	115	139	288.2	65.2	118.3	289.6
D	0	3.0	387.6	303.1	136	118	286.5	65.0	119.4	297.0
E	7.5	30.8	364.2	241.0	144	123	286.6	58.8	107.1	236.6
F	150.2	9.3	157.2	193.2	470	127	286.8	54.5	96.7	148.3
G	90.1	168.0	361.5	295.6	127	116	285.7	64.9	117.6	279.6
H	2.1	6.2	395.1	207.3	218	114	285.7	54.4	98.9	229.6

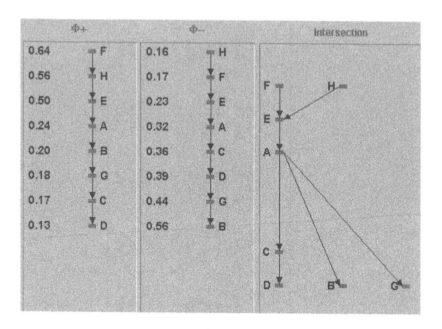

Figure 9.11 Júcar water supply system: Ranking of alternatives.

The ranking of alternatives resulting from the application of NAIADE impact analysis is depicted in Figure 9.11. From the figure, it can be inferred that:

1. The alternatives with the highest ranking are F (modernization of irrigation network Acequia Real del Júcar) and H (electrification of the remaining drought wells).
2. Lower rankings characterize alternatives E (wastewater reuse), A (use of the drought wells that are already electrified during drought situations), and C (systematic use of drought wells during winter).
3. Alternatives G (seawater desalination for the supply to Marina Baja), D (conjunctive application of all the short-term measures), and B (application of restrictions to the irrigation demands) exhibit the lowest ranking.

Results summarized by the impact matrix enable stakeholders to judge alternatives, according to their own preferences with respect to each assessment criterion, which allows filling in the preferences matrix. For the conflict analysis, the alternatives with lowest ranking have been discarded (D, B, and G). By applying NAIADE, the dendrogram of coalitions among stakeholders reported in Figure 9.12 was obtained. It can be inferred that:

- There is a high level of agreement among the users of irrigation water at Acequia Real del Júcar (Ribera Alta) and Canal Júcar-Turia.

- The following coalition includes the city of Valencia and the environmentalist groups. The irrigation users of Ribera Baja del Júcar are also included in this group.
- These groups together form a strong association against the hydroelectric users and the tourist water use board of the Marina Baja.

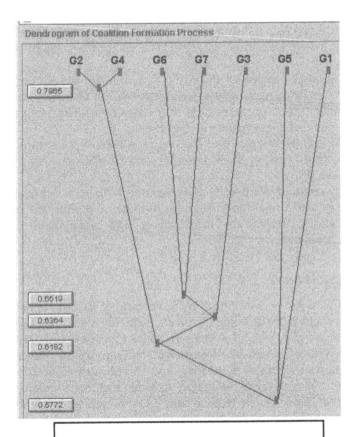

Stakeholders
G1: Tourist board of Marina Baja
G2: Farmers of Acequia Real del Júcar
G3: Farmers of Ribera Baja del Júcar
G4: Farmers of canal Júcar-Turia
G5: Hydroelectric company Iberdrola
G6: Environmental organizations
G7: Municipal users (City of Valencia)

Figure 9.12 Júcar water supply system: Dendrogram of coalitions among the different stakeholders.

In particular, the alternative accepted by all of the stakeholders is the modernization of Acequia Real del Júcar (alternative F). Partial consensus can be reached between different groups:

- The group formed by the irrigation farmers of Acequia Real and Canal Júcar-Turia and the group formed by the irrigation farmers of Ribera Baja, the city of Valencia, and environmentalists prefer alternative F (modernization of Acequia Real del Júcar) and in second place alternative E (wastewater reuse), vetoing alternatives A (use of electrified wells) and H (electrification of the remaining drought wells).
- The group formed by the city of Valencia and the environmentalist groups, together with the irrigation farmers of Ribera Baja, form a coalition with preference for alternative E (wastewater reuse), vetoing alternative A (use of drought wells).
- The city of Valencia and the environmental representatives form a coalition based on their preference for the alternative of wastewater reuse (E).
- The coalition formed by the irrigation farmers at Acequia Real del Júcar and Canal Júcar-Turia, prefers the modernization of Acequia Real del Júcar and the electrification of the remaining drought wells at Ribera Alta.

9.4 Conclusion

Adoption of a pro-active approach for drought mitigation is being recognized more and more as a key factor to effectively reduce the worst consequences of droughts, as well as to promote the efficient use of existing water resources, within the framework of sustainability principles. The preliminary identification and analysis of the long- and short-term measures to be implemented within a drought mitigation strategy requires software tools efficiently integrated within a decision support system in order to enhance their capabilities, as well as to simplify their use by decision makers. Further, the assessment of the appropriate mix of long- and short-term measures to cope with droughts should take into account different economic, environmental, and social criteria, as well as the preferences of the stakeholders affected by the decisions. Therefore multicriteria analysis represents the natural choice to perform such an assessment.

A procedure for the identification and assessment of long- and short-term measures for coping with drought in a water supply system has been presented. The procedure makes use of the simulation tool SIMGES for the simulation of the system and for assessing the effects of the different measures on the system performances. The multicriteria analysis technique NAIADE is then used to assess and rank the different alternatives based on a set of economic, environmental, and social criteria.

Examples of application to case studies in Italy (Simeto river basin water supply system in Sicily and Flumendosa–Campidano water supply system in Sardinia) and Spain (Júcar water supply system in Valencia region) are discussed in some details. Results of the applications show that the use of DSS and MCA can effectively improve the selection process of drought mitigation strategies in complex water supply systems under water scarcity threats.

Although the studies presented here have not been officially commissioned by the agencies responsible for the water systems management, the research cannot be considered just an academic exercise, thanks to the continuous contacts between such agencies and the university departments involved in the activities. Such contacts have concerned two crucial parts of the research: in a preliminary stage, the acquisition of basic data on water facilities features and on the drought mitigation measures proposed by the these agencies or by the government (in the Italian case by Sicilian Regional Drought Emergency Committee); in the final stage of the study, the definition of the values assigned to the criteria of interest for the agencies and a cooperation in the analysis of the resulting ranking of the alternatives. In light of this, one of the significant outcomes of the research activities can be considered the partial defeat of what is generally recognized as one of the major limits of the applicability of system analysis methodologies to real water resources systems, namely the gap between theory and practice.

9.5   Acknowledgments

This research has been carried out with the financial support of the European Commission program INCO.MED, within the project contract ICA3-CT- 1999-00014. Partial financial support by the GNDCI, U.O. 1.12, contract C.N.R. 01.01070.42, is also acknowledged. The application of the methodology to the Sicilian case study has been carried out with the help, in the preliminary stage, of engineers G. Parisi, G. Musumeci, and C. Di Bartolo. The Sardinian case study was based upon a study by the staff of Cagliari University, led by professor G. Sechi; the Spanish case study was based upon the study by the staff of the Universidad Politécnica de Valencia, led by professor J. Andreu.

References

Andreu, J., Capilla, J., and Sanchis, E. (1996). Aquatool: A generalized decision support system for water-resources planning and operational management. *J. Hydrology* 177, 269–291.

Andreu, J., Solera, A., and Sanchis, E. (2001, June 6–8). *Decision support systems for integrated water resources planning and management.* Proceedings of the Conference: Management of Northern River Basins, Oulu, Finland.

Cancelliere, A., Ancarani, A., and Rossi, G. (1998). Susceptibility of water supply reservoirs to drought conditions. *J. Hydrologic Eng.* 3(2), 140–148.

Duckstein, L. (1983). Trade-offs between drought mitigation measures. In V. Yevjevich et al. (Eds.), *Coping with droughts*. Littleton, CO: Water Resources Publications.

Dziegielewski, B. (2003). Long-term and short-term measures for coping with drought. In G. Rossi et al. (Eds.), *Tools for drought mitigation in Mediterranean regions* (pp. 319–339). Dordrecht: Kluwer Academic Publishers.

Flug, M., and Scott, J. F. (2000). Multicriteria decision analysis applied to Glen Canyon dam. *J. Water Resour. Planning Man.* 126(5), 270–276.

Goicoechea, A., Hansen, D. R., and Duckstein, L. (1982). *Multiobjective decision analysis with engineering and business application*. New York: John Wiley.

Loucks, D. P., and Da Costa, J. R. (1991). *Decision support systems, water resources planning*. Berlin: Springer-Verlag.

Loucks, D. P., and Gladwell, J. S. (1999). *Sustainability criteria for water resources system*. Cambridge, UK: Cambridge University Press.

Munda, G. (1995). *Multicriteria evaluation in a fuzzy environment. Theory and applications in ecological economics* (pp. 93–191). Heidelberg: Physica-Verlag.

Munda, G., Nijkamp, P., and Rietveld, P. (1994). Multicriteria evaluation in environmental management: Why and how. In M. Paruccini (Ed.), *Applying multiple criteria aid for decision to environmental management* (pp. 1–22). Dordrecht: Kluwer Academic Publishers.

Munda, G., Parruccini, M., and Rossi, G. (1998). Multicriteria evaluation methods in renewable resource management: Integrated water management under drought conditions. In E. Beinat and P. Nijkamp (Eds.), *Multicriteria analysis for land-use management* (pp. 79–93). Dordrecht: Kluwer Academic Publishers.

NAIADE Manual. (1996). *Reference guide version 1.0. ENG*, Joint Research Centre of the European Commission, ISPRA Site.

Raju, K. S., and Pillai, C. R. S. (1999). Multicriterion decision making in river basin planning and development. *Eur. J. Operational Res.* 112, 249–257.

Reitsma, R. F., Zagona, E. A., Chapra, S. C., and Strzepek, K. M. (1996). Decision support systems (DSS) for water resources management. In L. W. Mays (Ed.), *Water resources handbook* (pp. 33.1–33.35). New York: McGraw-Hill.

Rossi, G. (1996). Risorse idriche e sviluppo sostenibile. In S. Indelicato and M. Moschetto (Eds.), *La gestione delle acque in Italia* (pp. 73–91). Cosenza: Editoriale BIOS.

Rossi, G. (2000). Drought mitigation measures: A comprehensive framework. In J. V. Vogt and F. Somma (Eds.), *Drought and drought mitigation in Europe* (pp. 233–246). Dordrecht: Kluwer Academic Publishers.

Simonovic, S. P. (1996) Decision support systems for sustainable management of water resources. 1. General Principles. *Water Int.* 21(4), 223–232.

Wilhite, D. A. (Ed.). (2000). *Drought: A global assessment*. Vol. II. London and New York: Routledge.

World Commission on Environment and Development (WCED). (1987). *Our common future*. Oxford: Oxford University Press.

Yevjevich, V., Da Cunha, L., and Vlachos, E. (1983). *Coping with droughts*. Littleton, CO: Water Resources Publications.

Index

241

X

9 780367 391904